Lecture Notes in Computer Science 11782

More information about this series at http://www.springer.com/series/8277

Zhigeng Pan · Adrian David Cheok ·
Wolfgang Müller · Mingmin Zhang (Eds.)

Transactions on Edutainment XVI

 Springer

Editors-in-Chief
Zhigeng Pan
Hangzhou Normal University
Hangzhou, China

Adrian David Cheok
Imagineering Institute
Nusajaya, Malaysia

Wolfgang Müller
University of Education
Weingarten, Germany

Guest Editor
Mingmin Zhang
Zhejiang University
Hangzhou, China

ISSN 0302-9743 ISSN 1611-3349 (electronic)
Lecture Notes in Computer Science
ISSN 1867-7207 ISSN 1867-7754 (electronic)
Transactions on Edutainment
ISBN 978-3-662-61509-6 ISBN 978-3-662-61510-2 (eBook)
https://doi.org/10.1007/978-3-662-61510-2

This Springer imprint is published by the registered company Springer-Verlag GmbH, DE part of Springer Nature.
The registered company address is: Heidelberger Platz 3, 14197 Berlin, Germany

Editorial

In this issue, we have five sections on different topics. In the first section on E-learning and online Apps there are five papers. In the first paper by Ming Li et al., the authors conduct a study examining the effects of a motivational strategies-based edutainment intervention on college students' motivation and engagement in College EFL courses. In the second paper by Liying Wang et al., the authors collect the dataset including cognitive, emotional, and behavioral states of 35 students in daily self-regulated and single observed MOOC online learning by means of investigation and experiment. They conclude that emotional engagement has more significant impact on academic achievement than that of behavioral engagement, but less than that of cognitive engagement. In the third paper by Wenqin Qu et al., the authors propose a virtual bracket-interaction method based on the LDNI and the point set matching. The results show that their software is very helpful for orthodontists to complete bracket positioning. In the fourth paper by Mingliang Cao et al., the authors summarize the technologies, tools, and application scenarios relating to virtual pottery, including modeling methods; interactive tools and their advantages/disadvantages; and future trends in interactive methods. In the fifth paper by Chenyang Cui, the author divides the attributes of an artwork into two categories: subjective and objective, and proposes a method of attribute annotation for artworks based on the visual perception of the searcher or audiences.

In the second section on Image and Graphics, there are five papers. In the first paper by Lu Ye et al., the authors introduce a practical and new features fusion structure named "Dual Path Network" for road semantic segmentation. In the second paper by Wei Ye et al., in order to monitor the aging of transformers and ensure the operational safety in substations, a practical detection system is designed for indoor substation transformers based on the analysis of audio signal. In the third paper by Jinyu Ren et al., based on the big data statistics of the dry bulk carrier over 1,000 tons navigating on the Yangtze River, the authors analyze the distribution of the transport capacity and distribution of shippage, and sum up the linear formula between the gross ton and dead weight. In the fourth paper by Wensheng Yan et al., the authors propose R-SENet for foliage recognition to accurately identify plant leaves. In the fifth paper by Yiping Bao at al., the greedy strategy is used to improve the density clustering algorithm. In order to further improve the efficiency of density clustering algorithm based on the greedy strategy, the greedy strategy is applied to mining hot spots of taxi passengers.

In the third section on VR and AR, there are six papers. In the first paper by Jiahui Liu et al., the authors explore the visualization of marine ranch in tourist industry and provide a three-dimensional, dynamic, and visual model for tourism development in the coastal cities. In the second paper by Rui Dai et al., the authors first introduce related works in flood simulation modeling, VR training, and interacting techniques in VR systems. Then they discuss the implementation of a virtual training system for flood security education based on a Unity3D engine. In the third paper by Yu Gao et al., the authors present a VR experience system based on the theme of human body and cold virus struggle. Plentiful functions of virtual system were discussed in detail. In the fourth paper by Wei Xuan et al., the study focuses on the feasibility and development of AR technology for application to pancreatic surgery, especially the role of AR technology in medical teaching and clinical practice. In the fifth paper by Yuekun Jing et al., the authors propose a method of real-time human motion detection and skeleton control based on the depth image. The results show that the motion capture system can recover the 3D skeleton motion of the captured real human body even better. In the sixth paper by Zhigeng Pan et al., the authors construct a virtual experiment simulation situation with virtual human guidance in experiment operation, intelligent algorithm assistance in experiment failure, and systematic evaluation at the end of experiment.

In the fourth section on Computer Vision and AI, there are six papers in total. In the first paper by Lu Ye et al., the authors propose a method of driver's distracted behavior detection based on channel attention convolution neural network to solve the driver's distracted attention. In the second paper by Jinhong Li et al., the authors achieve data acquisition, motion feature extraction, and recognition of human motion by using motion capture device. In the third paper by Lilong Chen et al., the authors design a marching roadmap system based on osgEarth. In Chen's paper, the system can customize the vector information, provide a highly customized route curve type, as well as route dynamic setting and adjustment. In the fourth paper by Feng Gao et al., the authors provide a method for acquiring digital evidence on Play Station 4 and give a detailed description of embedded instant messenger software. In the fifth paper by Jia Wang et al., the study examines the mental health of depressed patients as the starting point and expands the communication experience of public service advertisements from the perspectives of psychology and behavior. In the sixth paper by Xiaoli Dong, she proposes some possible methods for making animation based on the study of national culture.

In the fifth section on Animation and Miscellaneous, there are five papers in total. In the first paper by Zhi-fa Du et al., it presents a fruit shape modeling method based on wavelet interpolation, trying to uniformly sample 2 to the power of m longitude lines of fruits with a upright axis, and uniformly sample 2 to the power of n points on each longitude line. In the second paper by Ying Zhang et al., the authors analyze partition, file system, relevant file, and other data of Xbox One concretely. In the third paper by Liying Wang et al., based on the dataset with 25 experimental participants, online learning engagement states are statistically analyzed and assessed by applying the schema including both prior rules and data fitting method which can quantitatively evaluate the learning engagement. In the fourth paper by Qing Cai et al., the authors propose a distributed cache and recovery method based on memory database to support systematic fault tolerance. In the fifth paper by Shen Jun, the communication mechanism of social media is explored, by analyzing the three core components: Platform, User, and Content of social media.

In addition, we would like to add some more words. The papers are selected from the workshop on Digital Heritage and Digital Museum, which was held during November 28–30, 2019, in Anqing, China. Furthermore, this issue will be the last one in the *Transactions on Edutainment Series*, based on the instructions from the publisher. A new journal will be created in the near future along the same lines as original one. A big thank you to Mr. Alfred Hofmann and Ms. Christine Reiss from Springer, and Miss Xinting Wang, my secretary from Hangzhou Normal University, for their kind efforts in helping prepare this journal.

February 2020   Zhigeng Pan
                                                                            Mingmin Zhang

Transactions on Edutainment

This journal subline serves as a forum for stimulating and disseminating innovative research ideas, theories, emerging technologies, empirical investigations, state-of-the-art methods, and tools in all the different genres of edutainment, such as game-based learning and serious games, interactive storytelling, virtual learning environments, VR-based education, and related fields. It covers aspects from educational and game theories, human–computer interaction, computer graphics, artificial intelligence, and systems design.

Daniel Thalmann   EPFL, Switzerland
Kok-Wai Wong   Murdoch University, Australia
Gangshan Wu   Nanjing University, China
Hyun Seung Yang   KAIST, South Korea
Xiaopeng Zhang   IA-CAS, China

Contents

VR/AR

CV and AI

Animation and Miscellaneous

E-learning and On-Line Apps

The Effects of Edutainment Strategies on Student Motivation and Engagement in College EFL Classes

Ming Li[1(✉)], Brett D. Jones[2], Gaigai Cao[3], and Dachao Wang[4]

[1] School of Foreign Languages, Shanghai University of Engineering Science, Shanghai, Songjiang 201620, China
bahalucy@163.com

[2] School of Education, Virginia Tech, Blacksburg, VA 24061, USA
brettdjones@gmail.com

[3] School of Statistics, Southwestern University of Finance and Economics, Chengdu 611130, Sichuan, China
philozhi@126.com

[4] School of Mechanical and Automotive Engineering, Shanghai University of Engineering Science, Shanghai, Songjiang 201620, China
bigchao@163.com

Abstract. This study examined the effects of a motivational strategies-based edutainment intervention on college students' motivation and engagement in College English as a Foreign Language (EFL) courses. The participants were junior students in the school of business who were divided into an experimental group and control group (N = 230). Independent samples t-tests, regressions, and correlations were used for data analysis. Activities such as the snowball activity and the gallery exhibition activity were used to stimulate students' situational interests. The results indicated that students in the experimental group had significantly higher motivation and engagement than those in the control group. Also, students' course perceptions predicted their engagement. Implications and limitations are discussed.

Keywords: Edutainment strategies · Motivation · Engagement · College English as a Foreign Language class

1 Introduction

Since the Reform and Opening up was initiated in 1978, China's higher education has been actively exploring how to stimulate students' interests and engage them in learning in order to obtain solid professional knowledge to meet the needs of the job market after they graduate. In early 2018, the implementation of the national standard of undergraduate education called for the establishment of student-centered classrooms [1]. Therefore, understanding how to stimulate students' interest has become the focus of college instructional reform among college professors. To address this need, we implemented an

Z. Pan et al. (Eds.): Transactions on Edutainment XVI, LNCS 11782, pp. 3–12, 2020.
https://doi.org/10.1007/978-3-662-61510-2_1

intervention that included some motivational strategies, including the snowball activity and the gallery exhibition activity, from the field of educational psychology to stimulate students' interests. The purpose of this study was to explore how the edutainment activities in this intervention influenced students' motivation and engagement in their learning. We believe that these strategies provide a practical approach to stimulating interest in college classes.

2 Literature Review

2.1 The MUSIC® Model of Motivation

Understanding how to motivate students to engage in learning has been a focus both in the field of educational psychology and in the field of English as a Foreign Language Teaching (EFL) teaching. As a result, a lot of research and theories relating to student motivation have been published and instructors have benefited. However, it can be hard for teachers and college professors outside the field of educational psychology to understand motivation research jargon and to implement the theories and strategies effectively. To address this problem, the MUSIC® Model of Motivation can be used by those who are unfamiliar with the terms in educational psychology [2]. MUSIC is an acronym for the five key words (eMpowerment, Usefulness, Success, Interest and Caring) that represent the major strategies teachers can use to motivate students. The MUSIC model is based on the research from many different theories, including self-determination theory (SDT) [3], future time perspective theory, expectancy-value theory, self-efficacy theory [4, 5], interest theory [6], attachment theory, and sense of community [7, 8].

In the United States, the MUSIC model has been implemented in many different level schools and it also was introduced to other countries such as Iran, Egypt, Colombia, and Iceland [9]. In China, the reliability and validity of the MUSIC model inventory was verified among the college students [10], and some of the strategies in the model were used as an intervention to increase student motivation and engagement and achievement [11, 12].

2.2 Student Engagement

Students' motivation is a key part of whether or not they will engage in their learning [2, 13]. In the realm of second language curriculum and instruction, Savignon emphasized that the essence and ultimate purpose of communicative language teaching was to help students engage in their language learning (p. 635) [15]. Obviously, China's College EFL class belongs to the realm of language acquisition; and therefore, it is true that the essence and purpose of college EFL class is student's engagement. Based on the internal relationships between the edutainment strategies in the MUSIC model and the essence and purpose of the college EFL class, it is reasonable and critical to implement the edutainment strategies into the college EFL class.

Student engagement has been defined as the student's active participation in academic and co-curricular or school-related activities, and commitment to educational goals and learning (p. 5) [13]. It is generally accepted that engagement can be divided into behavioral, emotional, and cognitive dimensions [14]. To examine changes in students' engagement after the use of edutainment strategies in the college EFL class, we focused on the behavioral and the cognitive dimensions of engagement.

3 Purpose, Research Questions, and Method

The purpose of this research was to test whether the use of the edutainment strategies related to the MUSIC model could increase students' motivation and engagement in college EFL classes. The research questions related to EFL classes were as follows:

RQ1: Is there a difference in students' motivation and engagement in the traditional lecture classes versus the classes incorporating the edutainment strategies from the MUSIC model?
RQ2: To what extent do college EFL students' motivation-related perceptions relate to their engagement?

This study was conducted in a university in Southeast China and the college EFL class was a compulsory one for all the college students in this university.

3.1 Participants

The participants were two college EFL teachers and their 115 students who were enrolled in the college EFL courses. The two EFL teachers had three years of experience in teaching this college EFL course. The entrance exam scores and the English test scores before the beginning of the first semester in the university were similar between the experimental group students and the control group students. Of the 245 students who agreed to complete the questionnaire, 230 students completed the online survey: 95.5% were male and 96% were Han nationality. The mean age of the sample was 18 years old.

3.2 Procedure

Study Design. Before beginning the research, the two instructors of the college EFL courses explained the purpose and the content of the research program to their students and asked them if they wanted to participate in the research. The ones who agreed to participate signed their name on a copy of the attendance list. The two teachers taught two classes respectively: one for the experimental group and the other for the control group. The research lasted eight weeks before the mid-term exam. There was one class meeting (for 90 min) each week and the students in the two groups used the same textbook. The only difference between the classes was that the instructors used the edutainment strategies in the experimental class while the instructors followed the traditional lecture style in the control group class. At the end of week eight, the instructors asked the participating students to complete an online questionnaire in class in order to examine their course perceptions.

The Experimental Group Classes. The two EFL teachers held a meeting one month before the semester began. They studied the relevant papers concerning the MUSIC model and chose the edutainment strategies to redesign their EFL course from the book entitled *Motivating Students by Design: Practical Strategies for Professors* [9], focusing on how to increase students' interests in the EFL course. For the experimental class, the instructors selected two activities from the book. One was the snowball activity and the other was the gallery exhibition activity. The details consisted of three steps. First, students were divided into small groups of six members according to their campus passport numbers. Then, the members of each group selected a leader among themselves. Second, the instructors gave students the rubric for their snowball activity and the gallery exhibition activity so that the members in each group would understand the expectations. Third, students participated in the snowball and gallery exhibition activities.

For the snowball activity, the instructor provided a task, such as the synonyms of a key CET 4 word, a 25-word summary of the text, and translation of a sentence in the exercise part in the textbook, and asked the students within one group to write down their answers on a piece of paper. After they finished the task, they crumpled the paper into a ball and threw it to another group in the class. The other groups did the same after they finished the task. Finally, the instructors unfolded the paper balls and assessed the answers on the sheets.

For the gallery exhibition activity, the instructor gave each group a large piece of paper (60 cm × 80 cm) and asked them to create their own concept map of the text. When each group finished with the map, the map was attached to the walls of the classroom. Next, the students walked around the classroom, read the map carefully, and the leader of each group drew stars on the map according to their group's overall evaluation of the map. In this way, the classrooms looked like a gallery as all the students walked around, talked, and marked a score on each concept map. Based on the mean score of the map, the class choose the best three and these three groups received some additional points.

The Control Group Classes. The control group classes followed traditional teaching approaches. The instructor read the new words, explained the meaning of them, and asked his or her students to read and memorize them. As for the content of the text, the instructors lectured using the PowerPoint slides associated with the textbook and emphasized the key points step by step. The instructors talked from the beginning to the

Fig. 1. The group members participating in the snowball activity

Fig. 2. The group members working on the gallery activity

end of the class. There were no interesting activities such as the snowball activity or the gallery exhibition activity in class and the instructors were always the leaders in class (Figs. 1 and 2).

3.3 Instruments

The MUSIC® Model of Academic Motivation Inventory. There are 26 items in the MUSIC Inventory and they belong to five scales titled empowerment, usefulness, success, interest, and caring. Each item is rated on a six-point Likert-format scale: 1 for *Strongly disagree*, 2 for *Disagree*, 3 for *Somewhat disagree*, 4 for *Somewhat agree*, 5 for *Agree*, and 6 for *Strongly agree*. The psychometric properties of the Chinese translation have been shown to be acceptable [10]. The complete inventory is available online [16].

Behavioral Engagement. Four items were used to measure students' effort on a six-point Likert-format scale:1 for *Strongly disagree*, 2 for *Disagree*, 3 for *Somewhat disagree*, 4 for *Somewhat agree*, 5 for *Agree*, and 6 for *Strongly agree*. In Li and Wang's research (2019), the reliability estimates were good ($\alpha = .82$).

Cognitive Engagement. Eight items were used to measure students' use of self-regulated strategies [17] on a six-point Likert-format scale: 1 for *Almost never*, 2 for *Seldom*, 3 for *Sometimes*, 4 for *Often*, 5 for *Almost always*, and 6 for *Always*. Prior studies documented that these eight items had acceptable reliability estimates ($\alpha = .90$) [11].

3.4 Analysis

We used three approaches to analyze the data, independent sample t-tests, general regression, and the Pearson correlation coefficients. The criterion level for statistical significance (p) was set to 0.05.

4 Results

4.1 Data Screening

After the implementation of the edutainment strategies in the college EFL classes for eight weeks, we collected 240 online questionnaires, 117 from the experimental group and 123 from the control group. To keep the same number of students in each group, we randomly selected 115 of the questionnaires from each group.

4.2 Internal Consistency

The Cronbach's alpha values were used to examine the internal consistency of the scales both for the MUSIC Inventory scales and for the engagement scales (see Table 1). The results revealed that both the MUSIC Inventory scales and the engagement scales had good to excellent reliability, with alphas ranging from .75 to .93.

Table 1. Reliability estimates, means, and standard deviations for the scales

Variable	α^a	α^b	M^a	SD^a	M^b	SD^b
eMpowerment	.84	.75	4.49	0.86	4.32	0.75
Usefulness	.88	.80	4.97	0.92	4.51	0.87
Success	.90	.86	4.17	1.07	3.78	1.00
Interest	.92	.88	4.60	0.89	3.55	0.88
Caring	.90	.79	5.24	0.72	4.83	0.59
Behavioral engagement	.84	.86	4.56	0.88	4.03	0.93
Cognitive engagement	.93	.88	3.62	1.03	3.14	0.79

[a] experimental group; [b] control group

4.3 Research Question 1

For RQ1 (Is there a difference in students' motivation and engagement in the traditional lecture classes versus the classes incorporating the edutainment strategies from the MUSIC model?), an independent t-test was conducted. Table 2 shows that there was

Table 2. Means and T-test results for the music model components and engagement components

Variable	M		Mean difference	t	p
	Experimental	Control			
Empowerment	4.49	4.32	0.17	1.58	0.11
Usefulness	4.97	4.51	0.46	3.81	**0.00**
Success	4.17	3.78	0.39	2.81	**0.05**
Interest	4.60	3.55	1.05	8.88	**0.00**
Caring	5.24	4.83	0.41	4.65	**0.00**
Behavioral engagement	4.56	4.03	0.53	4.48	**0.00**
Cognitive engagement	3.62	3.14	0.48	3.91	**0.00**

Note. Values in bold were statistically significant at an alpha value of 0.05.

a statistically significant difference between students' motivation (as indicated by the MUSIC model scales) and engagement in the two groups.

The results in Table 2 demonstrate that besides empowerment, the students in the experimental group reported higher levels of usefulness, success, interest, and caring than those in the traditional lecture class. As for behavioral and cognitive engagement, students in the experimental group devoted more effort to their EFL course and used more self-regulated strategies than those in the control group.

4.4 Research Question 2

The second research question was: To what extent do college EFL students' motivation-related perceptions relate to their engagement? We first calculated the Pearson correlation coefficients among the seven variables (the five MUSIC model variables and the two engagement variables). Table 3 shows that the seven variables were significantly correlated with each other except for the caring and success variables in the control group.

Table 3. Correlations among the study variables

Variable	1	2	3	4	5	6	7
1. Empowerment		$.50^{**}$	$.54^{**}$	$.27^{**}$	$.42^{**}$	$.30^{**}$	$.54^{**}$
2. Usefulness	$.49^{**}$		$.41^{**}$	$.49^{**}$	$.45^{**}$	$.47^{**}$	$.37^{**}$
3. Success	$.71^{**}$	$.59^{**}$		$.30^{**}$	$.13$	$.34^{**}$	$.61^{**}$
4. Interest	$.66^{**}$	$.68^{**}$	$.69^{**}$		$.54^{**}$	$.61^{**}$	$.47^{**}$
5. Caring	$.55^{**}$	$.59^{**}$	$.47^{**}$	$.74^{**}$		$.51^{**}$	$.39^{**}$
6. Behavioral engagement	$.58^{**}$	$.68^{**}$	$.64^{**}$	$.65^{**}$	$.53^{**}$		$.57^{**}$
7. Cognitive engagement	$.64^{**}$	$.50^{**}$	$.64^{**}$	$.65^{**}$	$.49^{**}$	$.74^{**}$	

Note. The correlations below the diagonal were for the experimental class and the correlations above the diagonal were for the control group.
* $p < 0.05$; ** $p < 0.01$

Next, the general regression analysis was conducted to examine the relationships between the five MUSIC model variables and behavioral and cognitive engagement. First, we examined the relationships between students' MUSIC perceptions and their behavioral engagement. In the experimental group, students' perceptions of usefulness and success predicted their behavioral engagement in the EFL class positively and significantly. In the control group, their behavioral engagement was predicted by the success, interest, and caring variables. The five MUSIC components explained 58% of the variance in students' behavioral engagement in the experimental group and 46% of the variance in students' behavioral engagement in the control group (see Table 4).

Table 4. General regression predicting behavioral engagement in the two groups

Predictor variable	R^2	df	ΔF	B	SE B	β[a]	t	p
	0.58[1]	114[1]	30.47[1]					
Usefulness[1]				0.36	0.09	0.38	4.25	**.000**
Success[1]				0.18	0.08	0.22	2.16	**.033**
	0.46[2]	114[2]	18.64[2]					
Success[2]				0.17	0.08	0.18	2.09	**.039**
Interest[2]				0.39	0.09	0.37	4.02	**.000**
Caring[2]				0.41	0.15	0.26	2.80	**.006**

Note: Values in bold were statistically significant at an alpha value of 0.05.
[1] experimental group; [2] control group

The results in Table 5 revealed that in both the experimental group and the control group the variables empowerment, success, and interest predicted students' use of self-regulated strategies positively and significantly. The five MUSIC components explained 52% of the variance in students' self-regulation in the experimental group and 53% of the variance in students' self-regulation in the control group (see Table 5).

Table 5. General regression predicting cognitive engagement in the two groups

Predictor variable	R^2	df	ΔF	B	SE B	β[a]	t	p
	0.52[1]	114[1]	23.55[1]					
Empowerment[1]				0.32	0.12	0.27	2.65	**.009**
Success[1]				0.21	0.10	0.22	2.01	**.047**
Interest[1]				0.33	0.15	0.29	2.25	**.026**
	0.53[2]	114[2]	24.11[2]					
Empowerment[2]				0.23	0.09	0.22	2.42	**.017**
Success[2]				0.35	0.06	0.45	5.32	**.000**
Interest[2]				0.22	0.07	0.74	2.83	**.005**

Note: Values in bold were statistically significant at an alpha value of 0.05.
[1] experimental group; [2] control group

5 Discussion and Conclusion

5.1 Research Question 1

The results indicated that compared to students in the control group, students in the experimental group found the course more useful, were more likely to believe that they could succeed, had more interest in the class, felt more cared for by their teacher, put forth more effort in the class, and used more self-regulated strategies in the class. These conclusions were consistent with that of the previous research conducted in other Chinese college EFL classes [11]. Therefore, it is reasonable for college EFL teachers to use the snowball activity and gallery exhibition activity to motivate students to engage in class.

Interestingly, the experimental and control group did not differ on their perceptions of empowerment. We cannot explain this finding because we expected students in the experimental group to report higher levels of empowerment given that they had more choices and autonomy. Further study is needed to understand this finding.

5.2 Research Question 2

The results of the correlational analysis showed that all of the MUSIC model components were related to behavioral and cognitive engagement. Furthermore, the results of the regression analyses showed that some of the MUSIC model components predicted students' behavioral and cognitive engagement in both the experimental and the control group. These results provide more evidence illustrating the theoretical relationship between the MUSIC model components and engagement [2]. Because the correlations indicated that the MUSIC model components were moderately related to each other, college EFL teachers who increase one of these variable may also increase one or more of the others, and thereby, increase students' engagement.

5.3 Limitations

First, this research was limited in scope because the sample included only two teachers and their two EFL classes. The results may be more generalizable if more EFL teachers with more students were included in the study. Second, most of the students were males who may differ from females in their perceptions of the class. Third, the intervention lasted only eight weeks. It is not known how long students' perceptions of these activities lasted throughout the course.

5.4 Conclusion

The edutainment strategies used in the intervention increased students' motivation-related perceptions of their EFL course (i.e., their perceptions related to usefulness, success, interest, and caring) and increased their behavioral and cognitive engagement in their learning. In addition, students' behavioral and cognitive engagement were predicted by some of the components of the MUSIC model. Thus, the strategies in the MUSIC model could be further used by the college EFL teachers in order to motivate their students engage in their learning.

Acknowledgments. The authors would like to thank Dr. Zhigeng Pan for his invaluable suggestions on this article. This paper was supported both by China's Foreign Language Textbooks and Pedagogy Research Center with the grant No. 2018SH0037A and by the project entitled First-class Discipline Construction of Foreign Language and literature at Shanghai University of Engineering Science with grant No. 2019XKZX011.

References

1. MOE: The national standards for quality teaching of undergraduates programs was issued (2018). http://www.moe.gov.cn/jyb_xwfb/xw_fbh/moe_2069/xwfbh_2018n/xwfb_20180130/mtbd/201801/t20180131_326074.html
2. Jones, B.D.: Motivating students to engage in learning: the MUSIC model of academic motivation. Int. J. Teach. Learn. High. Educ. **21**(2), 272–285 (2009)
3. Ryan, R.M., Deci, E.L.: Self-determination theory and the facilitation of intrinsic motivation, social development, and well-being. Am. Psychol. **55**(1), 68–78 (2000)
4. Bandura, A.: Social Foundations of Thought and Action: A Social Cognitive Theory. Prentice-Hall, Englewood Cliffs (1986)
5. Wigfield, A., Eccles, J.S.: Expectancy–value theory of achievement motivation. Contemp. Educ. Psychol. **25**(1), 68–81 (2000)
6. Hidi, S., Renninger, K.A.: The four-phase model of interest development. Educ. Psychol. **41**(2), 111–127 (2006)
7. Deci, E.L., Ryan, R.M.: A motivational approach to self: integration in personality. In: Nebraska Symposium on Motivation, vol. 38, pp. 237–288. University of Nebraska (Lincoln campus). Department of Psychology (1991)
8. Noddings, N.: The Challenge to Care in Schools: An Alternative Approach to Education. Teachers College Press, New York (1992)
9. Jones, B.D.: Motivating Students by Design: Practical Strategies for Professors. CreateSpace, Charleston (2018)
10. Jones, B.D., Li, M., Cruz, J.M.: A cross-cultural validation of the MUSIC® model of academic motivation inventory: evidence from Chinese- and Spanish-speaking university students. Int. J. Educ. Psychol. **6**(1), 366–385 (2017)
11. Li, M., Wang, S.: Examining the relationships between college students academic motivation, engagement and achievement under the guidance of the national standard of college education quality. Overseas English **6**, 9–12 (2019)
12. Li, M., Yu, L., Qin, Y., Lu, P., Zhang, X.: College student academic motivation and engagement in the college English course. Theory Pract. Lang. Stud. **6**(9), 1767–1773 (2016)
13. Christenson, S.L., Reschly, A.L., Wylie, C. (eds.): Handbook of Research on Student Engagement. Springer, Boston (2012). https://doi.org/10.1007/978-1-4614-2018-7
14. Fredricks, J.A., Blumenfeld, P.C., Paris, A.H.: School engagement: potential of the concept, state of the evidence. Rev. Educ. Res. **74**(1), 59–109 (2004)
15. Savignon, S.J.: Communicative language teaching: strategies and goals. In: Hinkel, E. (ed.) Handbook of Research in Second Language Teaching and Learning, vol. 1, pp. 635–651. Erlbaum Associates, Mahwah (2005)
16. Jones, B.D.: User guide for assessing the components of the MUSIC® model of motivation (2017). http://www.theMUSICmodel.com
17. Shell, D.F., Husman, J.: The multivariate dimensionality of personal control and future time perspective beliefs in achievement and self-regulation. Contemp. Educ. Psychol. **26**(4), 481–506 (2001)

Correlation Analysis Between Emotional Engagement and Achievement of University Students in Online Learning Based on an Elective Course

Liying Wang[1]([✉]) and Jingming Sui[2]

[1] School of Education Science, Nanjing Normal University, Nanjing 210097, China
wangliying@njnu.edu.cn
[2] Xinqiao School, Songjiang District, Shanghai 201612, China
1962276512@qq.com

Abstract. Emotional support in Online Learning (OL) has become a key factor in the success of self-regulated learning. Based on an elective course in the university, the dataset that includes cognitive, emotional and behavioral states of 35 students in daily self-regulated and single observed MOOC online learning are collected by means of investigation and experiment. The correlations between emotional engagement and achievement of online learning are analyzed via SPSS tool. The experimental data show that emotional engagement has the more significant impact on academic achievement than that of behavioral engagement, but less than that of cognitive engagement. The correlation between emotional engagement and cognitive engagement is extremely obvious, while the correlation between emotional engagement and behavioral engagement is extremely insignificant. Therefore, it is suggested that teachers could improve students' positive emotions and execution ability by establishing appropriate evaluation of curriculum value and self-perception to guarantee learning effect.

Keywords: Online learning · Emotional engagement · Cognitive engagement · Academic achievement · Correlation analysis

With more and more extensive and profound application of internet in global university, online learning mode has become an indispensable part of higher education students' learning process. Research on influencing factors of online learning is an important basis for predicting learning effect. Emotion is any conscious experience and emotions produce different physiological, behavioral and cognitive changes [1]. Positive emotions help broaden thinking, negative emotions help focus, and neutral emotions help calm analysis [2]. The multi-dimensional correlation analysis between emotional engagement and achievement of online learning, will promote the understanding of the influencing factors of learning effect, promote the humanized development of online learning, and enhance the interest and confidence of learning engagement in online learning.

© Springer-Verlag GmbH Germany, part of Springer Nature 2020
Z. Pan et al. (Eds.): Transactions on Edutainment XVI, LNCS 11782, pp. 13–24, 2020.
https://doi.org/10.1007/978-3-662-61510-2_2

1 Research on Online Learning Engagement

Schaufeli defined engagement as a positive, fulfilling, and work-related state of mind that is characterized by vigor, dedication, and absorption [3]. Fredricks et al. [4] proposed school engagement has the multifaceted nature which is defined in three ways of behavioral engagement, emotional engagement and cognitive engagement.

Garrison et al. [5] argued that metacognition is a higher-order, executive process that monitors and coordinates other cognitive processes engaged during learning. Garrison analyzed online collaborative inquiry discussions using the metacognitive construct which emphasizes self-regulation and self-efficacy process. Reeve et al. [6] proposed agentic engagement which emphasizes the constructive contribution of the students into the flow of the instruction they receive. Dixson [7] discovered in online classes multiple communication channels are clearly strongly correlated with higher student engagement.

Yin [8] discussed that emotional existence is the externalization of various emotions in the process of interaction between learners and online learning support technology tools, curriculum content, peers and teachers, and is the concrete expression of emotional engagement. Kahu [9] focused on two key elements of emotional engagement: interest and belonging to improve distance learning experience. Lee et al. [10] suggested educators and designers of online learning employ autonomy support strategies to engage students in active participation and successful completion of the course, because affectively engaged students feel satisfied from their achievement, enjoy interesting activities, and maintain a sense of self-worth in peer interactions.

Chinese research literatures mainly focus on learning engagement structure model building, since cultural differences will affect the result of emotional perception [11]. Zhu [12] gave a conceptual model of the relationship between the influencing factors of college students engagement, and quantitatively analyzed that college students' personal factors are the biggest influencing factors. Wei [13] developed emotion recognition module to enhance emotional interaction in online learning. Gao [14] drew a conclusion from the perspective of self-determination theory that the overall level of online learning engagement is not high, especially the level of emotional engagement needs to be improved. Gao [15] believed that online learning emotion is an important psychological factor affecting online learning engagement, in addition learning efficacy plays a completely mediating role in the process of positive emotion affecting online learning engagement. Therefore, students can improve online learning engagement by maintaining positive emotions and enhancing online learning efficacy. Liu [16] argued that online learners' perceived teacher support and their own learning engagement were not high in general, and that teacher support had a significant positive effect on online learning engagement. Yin [17] established a structural equation model of behavioral, cognitive, emotional and social interaction engagement, and considered that emotional engagement had a direct and significant positive impact on cognitive engagement. Ma [18] believed that cognitive competence and strategy, arousal and interest in metacognitive and emotional engagement factors significantly affect social learning behavior, while online discussion behavior significantly affects satisfaction and intention to continue learning curriculum.

In summary, students' online learning engagement level is not high, especially the emotional engagement level is obviously insufficient. Students' autonomous learning ability is not sufficient and need the support of teachers.

The rest of the paper is structured as below. Section 2 introduces the related theories and scales about academic emotion. Section 3 designs the framework of the correlation analysis between emotional engagement and achievement of university students in online learning, including our dataset collection schema based on an elective course. Afterwards, the correlation analysis results are clarified in detail. Finally, the conclusions are exhibited in Sect. 5.

2   Related Academic Emotion Theories and Scales

This paper applies the following psychological theories and scales to perceive the basic academic emotional engagement of the students.

2.1   Related Three Theories

Self-determination theory was put forward by Dexi and Ryan, emphasizing the active role of self in the process of motivation, and believing that motivation, internal needs and emotions are the motivation sources of self-determination.

Bandura's "social cognitive" theory holds that learning environment, self-awareness and learning behavior interact, among which self-efficacy plays an important role, and that physiological and emotional arousal state is one of the four possible sources of self-efficacy.

Fredrickson's "broaden and building theory of positive emotions" in 2001 holds that positive emotions benefit development and evolution for individuals and society. The differences between online learning and traditional learning in learners' emotional experience indicate that emotional engagement has an important impact on online learning effect. It is believed that positive emotions are more conducive to the emergence of positive learning behavior than neutral and negative emotions.

2.2   Related Three Scales

Scholars have investigated the subjective experience of learning emotions through questionnaires. Adolescent academic emotion questionnaire encoded by Dong et al. [19] has good reliability and validity. It is divided into four parts: positive high arousal, positive low arousal, negative high arousal and negative low arousal. Each part contains three to five specific emotions.

The academic self-efficacy scale compiled by Liang Yusong et al. in 2000 based on Pintrich and DeGroot's questionnaire, is divided into two dimensions: self-efficacy of learning ability (LA) and learning behavior (LB). The self-efficacy of learning ability refers to an individual's assessment of whether he or she is able to complete his or her studies, achieve good results and avoid academic failure. The self-efficacy of learning behavior refers to the students' assessment of whether their learning behavior can achieve their learning goals and the results of their own behavior. Each dimension contains 11 questions.

The achievement motivation is a tendency to pursue high standards. The AMS Achievement Motivation Scale was developed by Norwegian psychologists Gjesme, T. and Nygard, R. in 1970 and is divided into four levels. Achievement motivation is divided into two factors: approachability and avoidance, namely, motivation to hope for success (MS) and motivation to avoid failure (MAF). Achievement motivation has an important impact on students' learning attitude, persistence, learning task choice and academic performance.

Because emotion is a psychological element, its expression and perception are varied and vague. The conclusion of online learning emotional perception is incomplete and uncertain. From the perspective of emotional perception in online learning, this paper aims at the learners of an elective course, studies the emotional engagement of college students' MOOC learning, analyses its correlations with academic achievement. We attempt to put forward appropriate conclusions that complies for the online learning situation for university students and course teachers.

3   Correlation Analysis Framework Between Emotional Engagement and Achievement in Online Learning

The emotional engagement level of college students and its impact on their academic achievement of online learning are analyzed using the following structural models as shown in Figs. 1 and 2.

Fig. 1. Correlation model between basic learning emotions and academic achievement

According to Fig. 1, there are four independent variables: positive emotion, negative emotion, self-efficacy and achievement motivation. Positive emotion and negative emotion are collectively called basic academic emotion, Self-efficacy and achievement motivation are collectively called students' internal drive. The dependent variable is academic achievement. Their influences on academic achievement in daily self-regulated online learning process are investigated by questionnaires of scales in Sect. 2.2.

According to Fig. 2, the correlations between the four variables of academic achievement, behavioral engagement, cognitive engagement and emotional engagement are investigated and analyzed through single observed online learning experiment.

The experimental subjects including 35 students are selected from the course "Network Security and Maintenance" in 2017 in the major of educational technology of the university. Finally the dataset were obtained by questionnaires and experiments.

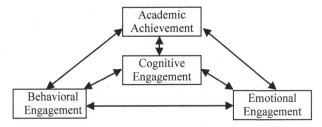

Fig. 2. Correlation model between learning engagement and achievement of online learning

The basic academic emotions, self-efficacy and achievement motivation of the subjects were investigated by the above scales in Sect. 2.2.

In the experimental online learning, the subjects were asked to complete their daily self-regulated and single observed online learning on the Chinese MOOC platform. The basic information of experimental online learning engagement of the subjects are obtained by questionnaires, camera and post-experiment test designed by ourselves.

Students are assigned to carry out self-regulated online learning for one semester. The selected MOOC course content of "Computer Network" of Harbin Polytechnic University [20] is closely related to the elective course. Finally, post-experiment tests matching the online learning content are released and recycled by rain classroom tools developed by Tsinghua University. All subjects are grouped to complete online learning, post-test and questionnaire in batches.

4   Correlations Between Emotional Engagement and Achievement

The recycled rate of this questionnaire is 100%. SPSS statistical data were used to analyze the correlations between all kinds of emotions and academic achievement.

4.1   Survey of Basic Learning Emotions and Achievement

The basic learning emotions are described in Table 1. It can be seen from the table that the average of positive high arousal (PHA) emotions and positive low arousal (PLA) emotions of subjects are significantly higher than the average of negative high arousal (NHA) and negative low arousal (NLA) emotions.

Table 1. Survey of basic learning emotions of students

Value	Type							
	PHA emotions	PLA emotions	NHA emotions	NLA emotions	LA self-efficacy	LB self-efficacy	MS motivation	MAF motivation
Mean	3.72	3.16	3.03	2.74	3.36	3.12	2.57	2.57

The mean of self-efficacy of Learning Ability (LA) is a little higher than that of Learning Behavior (LB). The Success of Motivation (MS) and Avoid Failure of Motivation (MAF) are balanced in values. It shows that the experimental subjects have relatively high learning enthusiasm for this study.

Academic achievement includes daily self-regulated and post-experiment test achievements shown in Table 2. From this, it can be seen that the proportion of students with more than 70 points is more than 89%, and two kinds of average scores are equivalent. Moreover, 14% of the students' daily self-regulated grades reaches a high level of over 90 points, and the average is higher than that of the post-experiment test. It indicates that some students' learning in the experimental environment might be affected a little.

Table 2. Survey of academic achievements in OL

Grade of daily self-regulated OL	Number of Students	Average score of daily self-regulated OL	Average score of post-experiment test
Bad: 40–50	4	46.38	45.00
Middle: 70–79	8	74.72	72.50
Good: 80–89	18	85.34	83.33
Excellent: 90–100	5	92.01	80.00

4.2 Correlations Between Basic Academic Emotions and Achievement

Various positive and negative emotions of students with same grade of daily self-regulated online learning (OL) are averaged as shown in Tables 3 and 4 respectively.

Table 3. Average positive emotions of students with same grade of daily self-regulated OL

Grade of daily self-regulated OL	PHA emotions			PLA emotions			Average of PHA emotions	Average of PLA emotions
	Pride	Enjoyment	Hope	Contentment	Calmness	Relief		
Bad	3.00	3.82	**4.31**	3.00	3.81	**2.75**	3.71	3.19
Middle	3.28	3.75	3.91	2.75	3.44	3.05	3.64	3.08
Good	3.34	3.71	4.00	3.04	3.50	2.91	3.68	3.15
Excellent	3.36	4.17	4.35	3.00	3.70	3.12	3.96	3.27

Table 3 shows the correlations between positive emotions and achievement as follows. (1) General tendency of the academic achievement of students increases with the improvement of positive emotions, through the average positive emotions of students with middle grade decrease a little. It can be seen that improving students' positive emotions is beneficial to improving students' academic achievement. (2) For the students with bad grade, "hope" is the emotion with the highest score and "Relief" emotion is the least. That means students with bad grade still have expectations for improving their academic achievement.

Table 4 shows the correlations between negative emotions and achievement as follows. (1) Negative emotions of most students are inversely proportional to the grade of their achievements. That means with the decreasing of negative emotions, students' scores of online learning will gradually improve. (2) From the view of the highest value of emotion in each grade of the academic achievement in OL, "anger" appears most frequently in students with bad academic achievement, and "depression" appears most frequently in other students. It can be seen that depression is the most common emotion among students and needs more attention. (3) "Anxiety" occurs more frequently in students with bad and middle grade, while students with good and excellent grade exhibit more balanced negative emotions. Therefore, the correlation between negative emotions and academic achievement is very high. It is necessary to reduce the negative emotions of the students appropriately and pertinently so as to keep their physical and mental health.

Table 4. Average negative emotions of students with same grade of daily self-regulated OL

Grade of daily self-regulated OL	NHA emotions			NLA emotions				Average of NHA emotions	Average of NLA emotions
	Anxiety	Shame	Anger	Boredom	Hopelessness	Depression	Fatigue		
Bad	3.68	2.95	**3.85**	2.00	2.30	3.80	2.56	3.49	2.67
Middle	3.36	3.05	3.23	2.57	2.88	**3.65**	3.19	3.21	3.07
Good	2.89	2.37	3.36	2.23	2.41	**3.39**	2.44	2.87	2.62
Excellent	2.91	2.48	3.52	1.91	2.64	**3.76**	2.45	2.97	2.69

4.3 Correlations Between Internal Drive and Achievement

The distribution of self-efficacy and achievement motivation with same grade of academic achievement are described in Table 5. The correlations between self-efficacy and achievement can be obtained from it as follows. (1) The average self-efficacy is greater

Table 5. Average self-efficacy of students with same grade of daily self-regulated MOOC OL

Grade of daily self-regulated OL	Average of LA self-efficacy	Average of LB self-efficacy	Value of LA-LB	Motivation value of MS-MAF
Bad	**3.91**	**3.32**	0.59	0.00
Middle	3.25	3.24	0.01	−0.27
Good	3.31	3.00	0.32	0.11
Excellent	3.27	3.22	0.05	0.00

than or equal to 3, which proves that most students have self-confident learning attitudes towards their abilities and behaviors. (2) Students with the bad grade have the highest self-efficacy which are far greater than that of students with other grades. That means they generally overestimate their achievements. (3) Most of the students whose grades are above 70 points are not blindly confident, but equivalent self-evaluation and more confident about their grades.

From Table 5 the following correlations between motivation and achievement can be obtained. (1) The positive difference between MS and MAF proves students have strong achievement motivation in pursuit of success. Students with good academic grade show that they are full of expectations for the improvement of their performance. (2) The negative difference between MS and MAF proves students have a stronger will to pursue improvement but fear of failure. Students with middle academic achievement grade have weak motivation for success though they want to further improve themselves but dare not make efforts. (3) The zero difference between MS and MAF proves students pursue success and fear of failure are relatively moderate equal. It occurs in students with bad and excellent grade of academic achievement.

4.4 Correlations Between Engagement and Academic Achievement in Daily Self-regulated Online Learning

The questionnaire to investigate daily self-regulated online learning engagement is divided into three parts with 19 questions, 7 questions are related to behavioral engagement such as learning time, 4 questions are related to emotional engagement such as learning satisfactory degree and trouble, and 8 questions are related of cognitive engagement such as learning importance and habits. Finally the quantitative statistical data are scored shown in Table 6.

Table 6. Survey of the investigation of daily self-regulated OL

Variables description	Average	Std. Variation	Max.	Min.	N
Behavioral engagement	16.943	3.7017	24	9	35
Cognitive engagement	15.457	2.1874	20	10	35
Emotional engagement	10.543	1.5782	14	7	35
Daily self-regulated OL score	62.9071	25.92080	90.59	0	35

The bivariate correlation between learning engagement and daily self-regulated OL score is analyzed by SPSS tool. The correlations can see from Table 7 as below.

Table 7. Correlations between engagement and achievement of daily self-regulated OL

Variables description		Daily self-regulated OL score	Behavioral engagement	Emotional engagement	Cognitive engagement
Daily self-regulated OL score	Pearson correlation	1	.053	−.089	.226
	Significant P value		.760	.612	.192
	N	35	35	35	35
Behavioral engagement	Pearson correlation	.053	1	.303	.432**
	Significant P value	.760		.077	**.010**
	N	35	35	35	35
Emotional engagement	Pearson correlation	−.089	.303	1	.531**
	Significant P value	.612	.077		**.001**
	N	35	35	35	35
Cognitive engagement	Pearson correlation	.226	.432**	.531**	1
	Significant P value	.192	**.010**	**.001**	
	N	35	35	35	35

**The significance P value below 0.01

From Table 7, these information can be drawn as following. (1) All significant P value between the four variables are greater than 0.05, which proves to some extent that the correlation between learning engagement and achievement are not obvious. (2) Comparatively, Pearson correlation between score and emotional engagement is greater than that of behavioral engagement, but is less than that of cognitive engagement. (3) Significant P value of cognitive and behavioral engagement is less than 0.05, indicating that there is a significant correlation between them. (4) Significant P value of cognitive and emotional engagement is 0.001, indicating that the correlation between them is extremely significant. (5) Significant P value of behavioral and emotional engagement is 0.077, which indicates that there is a moderate correlation between them. Therefore, three types of learning engagement are almost pairwise correlated.

4.5 Correlations Between Engagement and Academic Achievement in Single Observed Online Learning

The questionnaire of learning engagement after single observed online learning experiment includes three parts to investigate behavioral, emotional, cognitive engagement of the students, each part is designed with two questions and post-experiment test includes five questions of choice type. The statistical data are shown in Table 8.

Table 8. Survey of the investigation of single observed OL

Variables description	Average	Std. Variation	Max.	Min.	N
Behavioral engagement	3.543	1.2448	6	1	35
Cognitive engagement	21.657	2.7753	28	15	35
Emotional engagement	9.029	1.5240	11	4	35
Post-experiment test score	76.0000	26.03165	100	20	35

The correlations between learning engagement and the achievement in single observed OL is analyzed by SPSS and shown in Table 9.

Table 9. Correlations between engagement and achievement of single observed OL

Variables description		Behavioral engagement	Cognitive engagement	Emotional engagement	Post-experiment test score
Behavioral engagement	Pearson correlation	1	.107	.023	−.221
	Significant P value		.542	.898	.201
	N	35	35	35	35
Cognitive engagement	Pearson correlation	.107	1	.475**	.542**
	Significant P value	.542		**.004**	**.001**
	N	35	35	35	35
Emotional engagement	Pearson correlation	.023	.475**	1	.329
	Significant P value	.898	**.004**		.054
	N	35	35	35	35
Post-experiment test score	Pearson correlation	−.221	.542**	.329	1
	Significant P value	.201	**.001**	.054	
	N	35	35	35	35

**The significance P value below 0.01

The results from Table 9 indicates the correlations of them as follows. (1) Significant P values of behavioral, cognitive and emotional engagement and test score are 0.201, 0.001 and 0.054 respectively. It can be seen that emotional engagement and learning achievement are relatively highly correlated, while the cognitive engagement has the highest, and behavioral engagement has the least correlation to the achievement. (2) Significant P value of cognitive and emotional engagement is 0.004, which shows that the correlation between them is extremely obvious. (3) Significant P value of behavioral and emotional engagement is 0.898, which shows that the correlation between them is the least obvious.

5 Conclusions

Through the above data analysis, the correlations between emotional engagement and achievement of online leaning are clarified. The following conclusions and suggestions are tried to put forward.

(1) "Hope", a positive emotion and "depression", a negative emotion are the most common emotions among students. This shows that every student has high expectations of excellent grades, especially those with poor grades. Therefore, appropriately improving positive emotions and reducing negative emotions are conducive to improving students' academic achievement.

(2) Students with bad academic achievement have higher self-efficacy.
This group of students often overestimated their learning ability and learning behavior. Excessive self-evaluation can not bring them excellent results, but will affect their pursuit of excellent results. The gap between expectation and reality may bring about more negative emotions, further affect their learning effect. Therefore, this group of students should change their self-perception and self-evaluation level through correctly understanding themselves and their shortcomings.

(3) Achievement motivation has a significant impact on students who are middle and upper middle grades. Therefore, it is conducive to improving their academic achievement through stimulating students to raise awareness of the course importance and value.

(4) Emotional engagement has the more significant impact on academic achievement, cognitive engagement has the most and, and behavior involvement has the least impact on achievement.
There is a consistent tend of correlations in both daily self-regulated and single observed online learning process. It is suggested that educators can improve their academic achievement by establishing students' cognition of themselves.

(5) The correlation between cognitive and emotional engagement is extremely obvious, while the correlation between behavioral and emotional engagement is extremely insignificant.

This shows that the emotional orientation of college students are determined by cognitive evaluation, and the execution of learning behavior is not strong. Therefore, it is suggested that teachers should promote the occurrence and internalization of learning behavior through embodied teaching.

Finally, the conclusions of this study may be not universal enough for online learners as a whole. They still need to be further verified by expanding the subjects scale and the times of the experiment in the future works, since the experimental subjects are limited and confined to the same major which may lead to some deviations for the results.

Acknowledgments. The paper is supported by the Key Project of Ministry of Education for the 13th 5-year Plan of National Education Science of China [Grant No. DIA170375].

References

1. Emotion [EB/OL]. https://en.wikipedia.org/wiki/Emotion
2. Xu, Y., Zhu, C., Liu, D.: Proper use emotions to improve learning effect. Newspaper of Social Science in China, 27 March 2018. The third page of psychology (2018)
3. Schaufeli, W.B., Martinez, I.M., Pinto, A.M., et al.: Burnout and engagement in university students: a cross-national study. J. Cross Cult. Psychol. **5**, 464–481 (2002)
4. Fredricks, J.A., Blumenfeld, P.C., Paris, A.H.: School engagement: potential of the concept, state of the evidence. Rev. Educ. Res. **1**, 59 (2004)
5. Akyol, Z., Garrison, D.R.: Assessing metacognition in an online community of inquiry. Internet High. Educ. **3**, 183–190 (2011)
6. Reeve, J., Tseng, C.M.: Agency as a fourth aspect of students' engagement during learning activities. Psychol. Contemp. Educ. **4**, 257–267 (2011)
7. Dixson, M.D.: Creating effective student engagement in online courses: what do students find engaging? J. Scholarsh. Teach. Learn. **2**, 1–13 (2010)
8. Yin, R., Xu, H.: The progress and prospects of online learning engagement research in foreign countries. Open Educ. Res. **3**, 89–97 (2016). (in Chinese)
9. Kahu, E.: Increasing the emotional engagement of first year mature-aged distance students: interest and belonging. Int. J. First Year High. Educ. **2**, 45–55 (2014)
10. Lee, E., Pate, J.A., Cozart, D.: Autonomy support for online students. TechTrends **4**, 54–61 (2015)
11. Zheng, L., Li, A., Li, J.: Emotion recognition based on different culture. Inform. Technol. Inform. **5**, 60–62+69 (2009). (in Chinese)
12. Zhu, H.: An empirical study on factors of college students study engagement in China—base on behavior engagement, affective engagement and cognitive engagement. High. Educ. Forum **4**, 36–40 (2014). (in Chinese)
13. Wei, R., Ding, Y., Zhang, L., et al.: Design and implementation about emotion recognition module of online learning system. Mod. Educ. Technol. **3**, 115–122 (2014). (in Chinese)
14. Gao, J.: The relationship between extrinsic motivation and online learning engagement: a self-determination theory perspective. e-Educ. Res. **10**, 64–69 (2016). (in Chinese)
15. Gao, J.: Impact of online academic emotion on learning engagement: a social cognitive perspective. Open Educ. Res. **2**, 89–95 (2016). (in Chinese)
16. Liu, B., Zhang, W., Liu, J.: Study on the effect of teacher support on online learners' learning engagement. e-Educ. Res. **11**, 63–68 (2017). (in Chinese)
17. Yin, R., Xu, H.: Construction of online learning engagement structural model: an empirical study based on structural equation model. Open Educ. Res. **4**, 101–111 (2017). (in Chinese)
18. Ma, Z., Su, S., Zhang, T.: Research on the E-learning behavior model based on the theory of learning engagement—taking the course of "the design and implementation of network teaching platform" as an example. Mod. Educ. Technol. **1**, 74–80 (2017). (in Chinese)
19. Dong, Y., Yu, G.: The development and application of an academic emotions questionnaire. Acta Psychol. Sinica **5**, 852–860 (2017). (in Chinese)
20. https://www.icourse163.org/learn/HIT-154005?tid=1002210011#/learn/announce

Interactive Virtual Bracket Positioning Method in Orthodontics

Wenqin Qu[1], Ran Fan[2(✉)], and Zhigeng Pan[1,3]

[1] DMI Research Center, Hangzhou Normal University, Hangzhou 310012, China
annyqu@qq.com, 443922077@qq.com
[2] School of Information Science and Engineering, Hangzhou Normal University, Yuhang, Hangzhou 311121, China
fanran1029@gmail.com
[3] NINED Digital Technology Co., Ltd., Guangzhou 510000, China

Abstract. In the process of orthodontics, accurate bracket positioning is a necessary condition to quickly obtain the ideal treatment result. We propose a virtual bracket-interaction method based on the Layered Depth-Normal Images (LDNI) and the point set matching. By means of mouse-interaction, the fast movement of the bracket on the tooth surface is realized. Specifically, LDNI is used to complete the sampling of the bottom point of the bracket in space coordinate system. Then the nearest-distance algorithm is used to find the corresponding point on the tooth surface. Finally, the position of the bracket on the tooth surface is determined by point set matching. In addition, in order to help the orthodontist to complete clinical correction, we design a virtual bracket interaction software to realize the positioning operation of the virtual bracket through mouse interaction. The results demonstrates that the software we developed is very helpful for orthodontists to complete bracket positioning.

Keywords: Digital orthodontics · Bracket positioning · LDNI · Collision detection · Bracket interaction

1 Introduction

Orthodontics mainly wears the appliance on the surface of the teeth to make the teeth move physiologically, thereby achieving the purpose of orthodontics. In the process of orthodontic treatment, patients not only have requirements on the aesthetics of the teeth, but more importantly, the pursuit of comfort. It is hoped that the orthodontic teeth can play their chewing function normally. With the rapid development of computer technology, combining computer-aided design (CAD) and stomatology has become an important method for orthodontic research [1–3]. Computer-assisted orthodontic system allows doctors to quickly obtain correction data for patients' teeth, and uses 3D visualization technology to simulate the various processes of clinical orthodontic treatment, assisting doctors in analyzing orthodontic schemes and predicting orthodontic results. At the same time, doctors and patients can exchange opinions according to the visual treatment plan, which provides an intuitive and effective channel for communication between doctors

© Springer-Verlag GmbH Germany, part of Springer Nature 2020
Z. Pan et al. (Eds.): Transactions on Edutainment XVI, LNCS 11782, pp. 25–34, 2020.
https://doi.org/10.1007/978-3-662-61510-2_3

and patients. Therefore, the research and development of digital treatment assistance system has very important research significance and application value for orthodontics.

The work to be performed by the digital treatment assistance system is to build a dental model and provide data support for the design of the treatment plan [4]. To build an accurate tooth model, we first need to scan the plaster model through optical methods to obtain three-dimensional dental jaw data, and then use digital geometric processing technology to separate the tooth and gum data, repair the missing parts of the teeth, and build a movable tooth model [5]. Based on above, the doctor designs a path for orthodontics.

Designing the orthodontic path of teeth usually involves various orthodontic measurements and diagnosis. After designing the orthodontic path, it is necessary to install an orthodontic bracket to fix the orthodontic position of the tooth. However, the prior art lacks a method for correctly fixing the dental bracket [6]. In order to solve the above problems, this paper proposes a virtual bracket interaction method that uses LDNI and point set matching to achieve rapid movement of the bracket on the tooth surface through mouse interaction, thereby accurately fixing the bracket position.

2   Virtual Bracket Interaction Method

2.1   Bracket Positioning

Bracket positioning is to help doctors quickly determine the location of bracket installation in the clinic. In the process of positioning, we first obtain the patient's dental jaw model with a laser scanner, then complete the positioning of the virtual bracket in the digital orthodontic platform software, then generate a transfer adhesive tray with a 3D printer, and finally transfer the bracket Adheres to the patient's mouth [7]. The flow chart of bracket positioning is shown in Fig. 1. Among them, the adjustment of the position of the virtual bracket in the digital orthodontic platform software can be manually adjusted by means of mouse interaction according to the experience of the doctor.

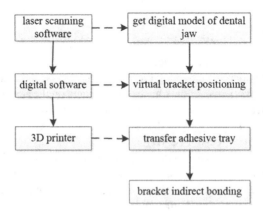

Fig. 1. Bracket positioning flowchart

Virtual bracket positioning is to help orthodontists perform bracket installation in the clinic. The difficulty in completing the bracket positioning is to rotate and translate the bracket to maintain the correct orthodontic force between the bracket and the teeth to achieve the effect of orthodontics. At the same time, during the movement of the bracket, no collision can occur between the teeth and the bracket, which requires us to add a collision detection function [8].

2.2 Interactive Method

This paper proposes a virtual bracket interaction method that uses LDNI and point set matching to achieve rapid movement of the bracket on the tooth surface through mouse interaction. Specifically, in the spatial coordinate system, LDNI is used to complete the sampling of the bottom point of the bracket, and then the nearest distance algorithm is used to find the corresponding point on the tooth surface, and then with the help of point set matching to determine the position of the bracket on the tooth surface. Here, we first introduce the concept of LDNI and its working principle, and then discuss the process of sampling the bottom point of the bracket using LDNI.

2.2.1 LDNI Introduction

LDNI is an extension of the B-rep form used to represent the entity model. It inherits the advantages of Boolean simplification, localization, and local coupling. It overcomes the shortcomings of B-rep in terms of computational strength by running algorithms directly on the GPU of the computer hardware. The boundary of the LDNI is represented by a polygon mesh, which can be obtained by fast sampling of computer hardware. The algorithms for the Boolean operations of the two LDNI solid models also run directly in the GPU efficiently. This method largely overcomes bottleneck of communication between graphics memory and main memory [9].

The LDNI of each solid model contains three directions of LDNI, each of which is perpendicular to the orthogonal axis (X, Y, Z axis). Assuming the sampling rate is w, the memory LDNI occupies is only $O(w^2)$, then use this advantage to achieve quick access to it. In addition, with the help of computer graphics hardware rasterization technology, we can complete the LDNI sampling based on the polygon mesh of the closed two-dimensional manifold.

2.2.2 LDNI Completes Sampling

The schematic diagram of the sampling process is shown in Fig. 2. It can be seen from Fig. 2 that the sampling is performed at a certain interval along the vertical direction and from the top to the bottom along the horizontal direction. Then, the gray values on each horizontal line are taken for one-dimensional scanning. Then, the one-dimensional scanning signal is sampled at a certain interval to obtain a discrete signal, that is, the entire sampling process is completed in the order of first sampling columns and then sampling rows. In this paper, we use LDNI to complete the sampling, and specifically use LDNI to process the points at the bottom of the bracket.

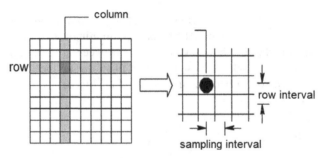

Fig. 2. Sampling diagram

From the schematic diagram of the two-dimensional sampling performed by LDNI in Fig. 2, we can see that in the process of sampling with LDNI, the sampling point is the intersection of the ray and the solid model [10]. The point is stored in a Hermite four-element structure, including a depth information that is d_x and the position information of the normal in three directions in space that is n_x, n_y, and n_z. After understanding the working principle of the LDNI, we use the Deep Peeling algorithm to complete the sampling, which is a technique that uses multiple passes to sort the depth values in a 3D scene [11]. In a standard depth detection scene, you only need to output the depth information of the layer closest to the viewpoint that is the shadow map to the screen. However, for a 3D solid model, it also has deeper layers such as the second layer and the third layer. This requires us to strip these depth information out in multiple passes.

2.3 Determine the Position of the Bracket on the Tooth Surface

Determining the position of the bracket on the surface of the tooth is actually a question of obtaining the spatial posture. The determination of the spatial pose can be converted into a problem of point set matching. After sampling the bottom point of the bracket with LDNI, we determined the position of the bracket on the surface of the tooth by finding the closest point that is the corresponding point on the surface of the tooth to the bottom of the bracket. In this paper, we use the Nearest-neighbor Search algorithm k-d tree to find the corresponding points.

3 Avoid Collision Between Bracket and Tooth

In the process of interaction between brackets and teeth, since brackets and teeth are rigid bodies, the phenomenon that brackets penetrate teeth cannot occur, so a collision avoidance function needs to be added. In the space domain, there are many methods for collision detection between rigid bodies. There are mainly two methods: space division and hierarchical bounding box [12]. Among them, the spatial division method has the problem of inconsistent detection efficiency, and the hierarchical bounding box method determines whether a collision occurs by judging whether an object is in contact with the constructed bounding box. In order to achieve accurate collision detection between brackets and teeth, this paper proposes to use LDNI, a special data structure, to determine the collision between teeth and brackets.

3.1 The Principle of LDNI to Achieve Collision Avoidance

We use LDNI's Boolean operations to detect collisions between teeth and brackets, and the definition of Boolean operation is shown in Table 1. For the two LDNI solid models A and B, if their projective lines overlap, the Boolean operation can be completed by sampling points with depth and normal information [13], which is easy to implement during the sampling process. Specifically, as long as the sampling frequencies of A and B are the same, their projective lines will overlap, and the efficiency of completing the Boolean operation will be high.

Table 1. Boolean operation definition

Operation object		Operator		
A	B	$\cup$	$\cap$	$\oplus$
T	T	T	T	F
T	F	T	F	T
F	T	T	F	T
F	F	F	F	T

In the process of detecting the collision between the bracket and the tooth, we also referred to previous research methods and used the technique of tactile rendering. Haptic rendering refers to using a haptic device to touch a computer-generated virtual object. It is mainly used for collision detection and distance calculation. VPS (Voxmap-PointShell) is a method for discrete representation of objects proposed by McNeely et al. [14]. The tactile bounding box records the intersection information of the discrete sampled light and the target object. When performing tactile calculation, as long as the line segment and bounding box intersection operation and five bilinear interpolation operations are performed once, collision information can be obtained at high speed [15]. Among them, the establishment of the tactile bounding box mainly includes two steps: (1) First construct a Position Bounding Box (Position Bounding Box) for the model, and get the position points of 6 surfaces; (2) Construct a DHC (Directional Hemi-Cube) for each position.

3.2 Discussion of Collision Avoidance

According to the International Dental Federation (Fédération Dentaire Internationale, FDI) position registration method, we give some instructions on the naming of teeth: each tooth is represented by a 3-digit symbol. The first symbol U/L represents the upper/lower teeth, the second symbol L/R represents the left/right teeth, and the third symbol represents the order of teeth. For example, UL7 indicates that the seventh tooth on the left side of the upper jaw is the second molar.

The effect of collision detection is shown in Fig. 3. Drag the bracket to any position on the UL7 surface of the tooth, and the bracket and the tooth surface remain in a fitted state in real time. The two will not collide. Then, the fit between the bracket and the teeth was observed from different angles.

(1) Effect of positive observation

(2) Effect of after adjusting viewing angle

Fig. 3. Collision detection during bracket movement

4   Virtual Bracket Interactive Software

In order to help orthodontic doctors to complete clinical correction, based on the above algorithmic ideas, this paper designs and develops a virtual bracket interaction software to realize the positioning of virtual brackets based on mouse interaction.

4.1   Software Basic Framework

The development platform of the virtual bracket interactive software is Microsoft Visual Studio 2013, and the development language is C++, and use OpenGL graphics library for model rendering. The software can run on Win7 operating system (X64), and there is no special requirement for the configuration of the computer itself.

The software interface mainly includes three functional modules: a toolbar, a dental information display panel, and a main view window. "Show upper jaw", "show lower jaw", "show gum", and "show teeth" in the toolbar are used to meet the needs of orthodontists for dental information of patients at different stages of clinical operations. For example, when the orthodontist is performing bracket mounting on the maxillary teeth, just click "show maxillary" to filter other irrelevant tooth information. At the same time, "Looking Up", "Looking Down", "Left", "Front", and "Right" help to quickly switch to the perspective that the orthodontist needs to observe, and you can also obtain the dental jaw models in different perspectives by moving the right mouse button.

4.2 Software Functions

Bracket adjustment is the most important function in the virtual bracket interactive software. The "Adjust" button includes two adjustment methods: quick positioning by dialog box and manual movement. Now, we take the adjustment of the position of the maxillary bracket as an example to introduce the function of the virtual bracket interactive software.

(1) Click "Open" and select the dental and jaw model data of the patient. The data reading interface is shown in Fig. 4.

Fig. 4. Access interface for dental jaw model data

(2) Click "Show gums" and "Show maxillas" to get the positional relationship of the maxillary teeth. Then, you can switch the viewing angle by moving the right mouse button. At the same time, you can drag the entire dental model by pressing the mouse wheel. Through certain mouse adjustment interactions, the three-dimensional maxillary tooth data in a specific perspective is obtained as shown in Fig. 5.

(3) Click the "Adjust" button to launch the bracket adjustment dialog shown in Fig. 6. For bracket movement, you can click the four directions (up, down, left and right) and follow the amount of single-step movement. For roll, you can click the up, down, left, and right directions for multiple times to complete the amount of rotation.

Fig. 5. Maxillary 3D dental data

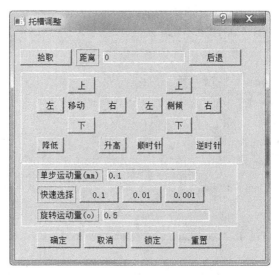

Fig. 6. Bracket adjustment dialog

By clicking the button in the bracket positioning box, you can directly see the movement of the bracket in the teeth, which greatly helps the orthodontist's clinical operation. The initial position of a tooth and the results after bracket adjustment are shown in Fig. 7.

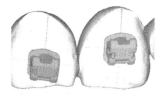

(1) Before bracket adjustment

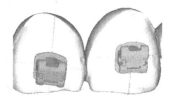

(2) After bracket adjustment

Fig. 7. Effect of before and after bracket adjustment

5 Conclusion

This paper provides a rapid interaction method for brackets positioning on the tooth surface. This method accurately detects collisions between teeth and brackets in real time. The method is efficient and stable so that doctors can design a correct treatment plan in an intuitive manner.

Of course, there is still some space for improvement. In the algorithm, after sampling the points at the bottom of the bracket, more points need to be found on the surface of the tooth, which results in a lower efficiency of the search process. Therefore, in further research, we need to find a more efficient point set matching algorithm.

Acknowledgements. The authors would like to thank the support of the Guangzhou Innovation and Entrepreneurship Leading Team Project under grant CXLJTD-201609.

References

1. Rekow, E.D., Erdman, A.G., Riley, D.R., et al.: CAD/CAM for dental restorations-some of the curious challenges. IEEE Trans. Biomed. Eng. **38**(4), 314–318 (1991)
2. Moeslund, T.B., Hilton, A., Krüger, V.: A survey of advances in vision-based human motion capture and analysis. CVIU **104**(2–3), 90–126 (2006)
3. Duret, F., Preston, J.D.: CAD/CAM imaging in dentistry. Curr. Opin. Dent. **1**(2), 150–154 (1991)
4. Li, C., Niu, J., Dai, D.: Application of digital correction technology in orthodontic treatment. Gen. J. Stomatol. **5**(33), 17–23 (2018)
5. Fan, R., Niu, Y., Jin, X.: Computer-assisted invisible orthodontic system. J. Comput.-Aided Des. Comput. Graph. **25**(01), 81–92 (2013)

6. Tang, W., Wang, T., Wang, D.: Research progress of orthodontic bracket positioning methods. J. Second Mil Med Univ. **37**(05), 613–617 (2016)

7. He, Y., Ji, L., Chen, F.: Application of digital virtual positioning in orthodontic bracket bonding training. Contin. Med. Educ. **33**(08), 22–24 (2019)

8. Peng, Y., Cheng, X., Dai, N., et al.: Research on collision detection in orthodontic simulation. Chin. Manuf. Informatiz. **38**(23), 19–24 + 28 (2009)

9. Wang, C.L., Leung, Y.S., Chen, Y.: Solid modeling of polyhedral objects by layered depth-normal images on the GPU. Comput.-Aided Des. **42**(6), 535–544 (2010)

10. McNeely, W.A., Puterbaugh, K.D., Troy, J.J.: Six degree-of-freedom haptic rendering using voxel sampling. In: Proceedings of the 26th Annual Conference on Computer Graphics and Interactive Techniques (1999)

11. Sun, M., Lv, W., Liu, X.: Approximate soft shadow algorithm combining deep stripping and GPU. J. Image Graph. **15**(09), 1391–1397 (2010)

12. Dan, K., Geng, G., Zhou, M.: Research on rigid body and soft collision detection algorithm in virtual surgery. Comput. Technol. Dev. (09), 60–63 (2008)

13. Wang, C.L., Manocha, D.: GPU-based offset surface computation using point samples. Comput.-Aided Des. **45**(2), 321–330 (2012)

14. Motohashi, N., Kuroda, T.: A 3D computer-aided design system applied to diagnosis and treatment planning in orthodontics and orthognathic surgery. Eur. J. Orthod. **21**, 263–274 (1999)

15. Han, X., Wan, H., Zhou, Z.: Real-time haptic rendering of highly complex pseudoconvex. J. Comput.-Aided Des. Comput. Graph. **21**(01), 60–66 (2009)

Research and Implementation of Virtual Pottery

Mingliang Cao[1,2(✉)], Pei Hu[3], and Mingtang Li[3]

[1] Guangdong Academy of Research on VR Industry, Foshan University, Foshan 528000, China
merlin.cao@connect.polyu.hk
[2] Guangzhou NINED Digital Technology Co., Ltd., Guangzhou 510000, China
[3] Automation College, Foshan University, Foshan 528000, China

Abstract. Virtual reality technology plays an increasing important role in the study and implementation of pottery. This paper mainly summarizes the technologies, tools and application scenarios relating to virtual pottery, including modeling methods; interactive tools and their advantages/disadvantages; future trends in interactive methods; and, applications of virtual pottery in museums, teaching, archaeology and entertainment.

Keywords: Interaction methods · Applications · Virtual pottery

1 Introduction

Pottery is a comprehensive art, which has been subject to a complex and long process of cultural accumulation. The term "pottery" refers to the production of clay vessels. Mastering the requisite skills for crafting pottery eludes most novices due to the need for extensive practice, which requires access to a suitably equipped pottery studio or many tutorials. With developments in society over time, the design of pottery has developed gradually from manual to computer-aided. Computer technology has changed the way pottery products are designed [1]. In order to meet people's needs, virtual reality technology is emerging at this historic moment. With the help of virtual reality technology, it is easy to modify a design repeatedly, present a very realistic effect and improve the production efficiency [2]. Nowadays, virtual reality technology is not only used in the production of pottery, but also in the restoration of broken pottery relics, making a significant contribution to the preservation of cultural heritage [3]. At the same time, virtual reality has helped to change how pottery relics are displayed. Instead of seeing the real work, people can observe it from all directions and angles through a virtual reality platform [4]. At present, the mature virtual reality technology has enabled the application of virtual pottery to be extended to education and entertainment [5].

The interactive mode is the most critical technology relating to virtual pottery. This paper summarizes the development of virtual pottery, compares the advantages and disadvantages of several interactive modes, introduces several application scenarios, and predicts some trends in future development trend.

© Springer-Verlag GmbH Germany, part of Springer Nature 2020
Z. Pan et al. (Eds.): Transactions on Edutainment XVI, LNCS 11782, pp. 35–44, 2020.
https://doi.org/10.1007/978-3-662-61510-2_4

2   Modeling and Interaction of Virtual Pottery

At present, the research relating to the making of virtual pottery focuses mainly on modeling methods and interactive modes.

2.1   Modeling of Virtual Pottery

Discussions of virtual pottery modeling methods mainly focus on geometric and physical modeling. Geometric modeling involves modeling from shape and appearance, while physical modeling is concerned with deformation and collision detection and the responses of objects (see Fig. 1).

Fig. 1.   Modeling methods of virtual pottery.

2.1.1   Geometric Modeling

As the most basic and important part of virtual reality technology, geometric modeling is the modeling of the shape and appearance of 3D graphics. In geometric modeling, points and lines are used to construct the outer boundaries of 3D objects, that is, only the boundaries are used to represent the 3D objects [6]. For example, Cho et al. [7] proposed a digital ceramic modeling system, in which users can create organic forms of ceramics that are difficult to achieve in the real world but exist in the virtual world. This is a natural, intuitive and interactive 3D modeling method. Murugappan et al. [8] proposed an interactive system called Handy-Potter, which is capable of creating a variety of constrained, free forms very quickly and a wide variety of shapes in seconds, allowing users to create, modify, and manipulate 3D shapes directly, enabling online creation and modification of generalized cylinders without extensive training. In another study, Vinayak et al. [9] added sculpting and smoothing tools to the geometric modeling process in order to create local collapses in the ceramics and help alleviate local irregularities.

2.1.2   Physical Modeling

The essence of physical modeling is to simplify and abstract the complex problems in real life, and design them into specific graphics by using the knowledge of physics and computer technology. For example, Lee et al. [10] proposed a virtual pottery modeling system, in which the clay is represented by a sector element according to the cylindrical symmetry of the pottery, and the virtual clay is deformed by tactile tools. Practice shows that the modeling system is efficient, intuitive and provides force feedback. Chaudhury and Chaudhuri [11] not only proposed a virtual pottery deformation model with simple and effective collision detection and response, but also used the Rayleigh density function to distribute the removed clay into the whole clay volume so as to retain the volume; this is close to the actual pottery production process and enhances the user's sense of immersion.

2.2 Interaction of Virtual Pottery

With developments in society, the ways of human computer interactions are changing constantly. At present, there are several popular ways of interacting in virtual pottery applications, including through a mouse, data gloves, depth camera and Leap Motion. Their advantages and disadvantages are shown in Table 1.

Table 1. Comparison of popular ways of interaction in virtual pottery

The type	Mouse	Data glove	Kinect and Leap Motion
Advantage	● It is easy to operate and cheap	● The field of view is not restricted; ● There is tactile feedback	● Low threshold and flexible scene; ● Users do not need to put the devices on and off their hands
Disadvantage	● Angle range is too small	● It has high delay, low resolution, and small range of motion; ● It's inconvenient to use	● The field of view is restricted; ● There's no tactile feedback

2.2.1 Mouse

In virtual reality, the mouse is the earliest form of interaction. People use a 2D mouse to draw on a single 3D image projection. Zheng [12] described first mapping the unearthed broken pottery into a virtual space, and then using the mouse to manipulate the object for virtual restoration. However, there is a serious problem by using a 2D mouse to control 3D objects. In the physical world, our hands naturally control the position of x, y, z, and three axes of rotation. The traditional 2D mouse can only directly affect two of these six variables at a time.

2.2.2 Data Glove

With the developments in science and technology, interaction using a mouse no longer met people's needs, and the next method was the data glove. Korida et al. [13] reported a study in which virtual objects were presented stereoscopically just in front of users through LCD shutter glasses, enabling them to manipulate the objects directly by using their own real hands with two-handed instrumented gloves. This provided functions such as dynamic gesture expressions, indications of the positions and sizes of the 3D objects, spatial manipulations, and a stereoscopic display function to operate with the two-handed gloves. A method called quasi-force-feedback was designed to control spatial tasks such as translations, stations and deformations of objects. This took advantage of an object's deformation speed to give the users a visual sense of its force resistance without using any real force-feedback devices. Han and Han [14] used OptiTrack (www.optitrack. com), which contains not only data gloves but also behavior trackers, to track gestures and create three-dimensional pottery objects and sound shapes. However, such a system cannot be accessed easily by ordinary users outside the laboratory environment, and wearing or holding the gloves may cause interference to users during the centralized modeling task (see Fig. 2).

Fig. 2. Using data gloves to interact with virtual pottery.

2.2.3 Kinect and Leap Motion

With the advent of depth cameras, it became possible to interact directly with bare hands (see Fig. 3). One is the interaction of predefined gestures. People need to learn the definitions of various gestures in advance, and then track the gestures with the depth camera. Predefined gesture interaction has been widely applied in virtual potteries. For example, Mustafa and Ismail [15] used Unity3D to integrate gesture functionality, then used Leap Motion (www.leapmotion.com) to track input gestures and sensors to calibrate them. However, using predefined gesture interaction has significant limitations. In the process of modeling, it can be a waste of users' time to learn too many gestures, but if they learn too few they cannot operate since, in the space creation expression, intention is not enough. The user should focus on the design task, rather than spend time on learning and memory signals. In order to solve this problem, Vinayak and Ramani [16] proposed a way to interact without requiring users to learn and remember gestures. Instead, it realized gesture capture through Soft Kinetic DS325 and Leap Motion controller, and then used an exponential smoothing method and selective Laplace smoothing method to make the shape of virtual ceramic mesh conform to the shape of the user's hand. Therefore, there was no need for any fixed gesture grasping and shape deformation, and the intention of deformation could be expressed directly by using hands and fingers. Park et al. [17] not only proposed to track gestures with the Leap Motion and Kinect camera, but also added a HMD virtual reality device to the interaction. Because this worked in the computer environment, the front and side of a piece of ceramic art could not be observed, but the 3D model could be observed through the HMD device. Neira's research [18] went one step further. He used Leap Motion to track the palm and interact with the environment. Two small tactile vibrators were added to each hand, allowing users to get tactile feedback and increase their sense of immersion.

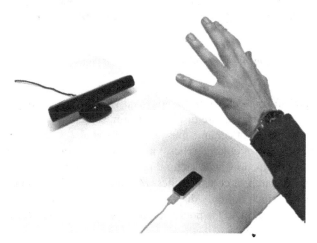

Fig. 3. Using Leap Motion and Kinect to interact.

3 Implementation of Virtual Pottery

At present, although virtual pottery is not widespread, it has been applied in many contexts (see Fig. 4).

3.1 Museum

Developments in computers have promoted developments in all walks of life. Museums have been able to display different forms of pottery from different periods. The early museums mainly displayed and studied objects representing natural and human cultural heritage. The emergence of computers has changed the traditional forms of museums, as they have gradually become digitized. The digital museums in our country are essentially two-dimensional. However, the two-dimensional digital museum is gradually becoming unable to meet people's needs, so the virtual reality museum has come into being. Huang et al. [19] realized the visual experience of a digital ceramic museum by using

Fig. 4. Application scene of virtual pottery.

multimedia design software, VRML technology and WEB2.0 technology. They created an interactive virtual ceramic culture display platform, something of great significance for protecting the resources of ceramic museum. Avella et al. [20] established a system that integrated the display of cultural relics, virtual view of debris recovery, and multiple uses of visitors' interactions with cultural relics, which provided great help for the understanding and research of ceramic relics. Martínez et al. [21] studied an application called TinajAR, which not only displayed pottery products in museums, but also added a virtual avatar to explain the pottery production process and investigated users' satisfaction.

3.2 Teaching

Virtual reality in teaching research, such as the early work of Moore [22], focuses on education theory, method and practice. Early on, Han et al. [23] proposed a pottery teaching system. They designed a tangible interface and integrated it into the real environment for interaction, and eventually provided six interaction modes. Chiang et al. [24] introduced a novel instructional system referred to as PotteryGo, which integrated sensor-based gesture recognition with 3D modeling techniques within a virtual reality environment. The deformations in the virtual vessels in the process were controlled using hand gestures detected by motion sensors. By analyzing the experiences of users with no prior pottery experience, some advantages were identified: (1) After receiving the tutorial, the participants were able to make a ceramic base using fewer attempts than before the tutorial; (2) Participants who received guidance through visual feedback required fewer attempts than those who received no visual guidance; and, (3) Participants who underwent this training were able to produce an actual ceramic base as easily as individuals who received instruction in a real-world pottery studio. In Wang's research [25], students were even allowed to learn and make pottery on iPads through the program potteryHD.

Using approaches such as those described here, students can learn and experience pottery simultaneously, which is not only convenient for the students but can also improve the production efficiency.

3.3 Archaeology

Virtual pottery has been studied extensively in archaeology. Cohen et al. [26] presented a method to assist in the tedious task of reconstructing ceramic vessels from unearthed archeological shards (fragments) using 3D computer vision-enabling technologies. The method exploited the shards surface intrinsic differential geometry information (one of many possible tools) coupled with a series of generic models to produce a virtual reconstruction and rendition of what the original vessel may have looked like. Iqbal et al. [27] used a virtual method of optical geometry to reconstruct objects from brittle fragments. This kind of technology relied on the broken surface boundaries and on their profiles in order to determine the correct matches among fragments. Pots could be modeled with their shapes as surfaces of revolutions, depending on their profile lines and axes of revolution. Kampel and Sablatnig [28] proposed a prototype system for the automatic archiving of archaeological fragments. The work was performed in a framework of documenting ceramic fragments. The methods proposed were tested on synthetic and real data with reasonably good results since they were better than traditional manual archaeological radius and volume estimation.

3.4 Entertainment

Life without entertainment would be tedious. Nowadays, many people go to studios to learn ceramics. The appearance of virtual ceramics does not require users to go to a special gallery to experience the process of making pottery [29]. Through the virtual reality method, the whole process of casting, glazing, firing and finished products can be realized in the virtual environment, and the finished work can be displayed for others to appreciate. This method makes the creation of pottery intuitive and simple, and enables more non-professionals to experience the real environment of pottery creation. Gao et al. [30] proposed an interactive virtual ceramic device, which provided a simple creation interface for users to enjoy making ceramic work. In addition, the finished work could be saved and shared automatically through social networks.

4 Discussion

In the interaction of virtual pottery art, the Leap Motion device is relatively convenient by using bare hands. Neira [18] increased tactile feedback during interaction, by adding two small phone vibrators in each hand to enable the user to feel the touch and clay deformation. However, because of the influence of the phone vibrator cable, the Leap Motion track sometimes disappears, so we suggest that in future studies: (1) It would be better to use wireless connections; (2) In the process of making pottery, the fingers will inevitably touch the clay. Tactile feedback should be added not only to the palm of the hand, but also to each finger, so as to make the immersion stronger.

When only hands are interacting in the environment, it will appear awkward. If a virtual character is added to the environment, the whole picture may appear more harmonious, and it's better to have a virtual character around to guide you. Makransky et al. [31] added virtual character guidance to their science lab simulations, and the results were better than they were without this guidance. Meanwhile, we can improve the visualization effects of virtual character by using both professional equipment to capture the body gestures and Leap Motion or data glove to capture the hand gestures.

Cho et al. [7] added a wooden rotating wheel in the real world. When he turned the wheel with one hand, the wheel in the virtual world also rotated, and then interacted with the virtual pottery with the other hand. This was very inconvenient. It would be better if we could add a virtual wheel directly to the virtual world and set a button that could adjust the speed manually. When we wear the data glove with tactile feedback, we can not only visually see the virtual wheel but also feel the touch of the virtual pottery at different speeds by hand.

In terms of virtual ceramic teaching, there is not much research in the existing literature; the research has only focused on design. There is little sense of virtual immersion. The use of VR headsets during teaching can increase students' sense of being immersed and make them feel like they are conducting experimental teaching in real life. Therefore, the addition of VR headsets is an important direction for our future research.

5   Conclusion

This paper has provided a critical review of the technology and applications of virtual pottery in the virtual reality field. Based on this review, the authors propose several potential research directions for virtual pottery technology including: (1) We can add wireless vibrators to future studies of tactile feedback in virtual ceramics; (2) We can add virtual avatars to guide virtual pottery study or application; (3) We can use both professional motion capture equipment and Leap Motion or data glove together to capture the gestures of body and hands for improving the visualization effects of virtual avatar. All in all, the most important thing is to improve the user's immersion and hand's authenticity in interactions when making or appreciating pottery virtually. Pottery technology has been passed on from generation to generation for thousands of years, and the integration of it into virtual reality is the current development direction. Therefore, this is the future of virtual pottery.

Acknowledgment. We would like to acknowledge the support of grant 2018KTSCX242. Also, we acknowledge the support of the Guangzhou Innovation and Entrepreneurship Leading Team Project under grant CXLJTD-201609.

References

1. Wang, A.H., Sai, S.T., Liu, Y.M.: The high computer technology application study about the daily-use ceramic products design. In: International Conference on Future Information Engineering, vol. 10, pp. 184–189 (2014)

2. Zhu, S.S., Li, S.H.: Redesign of Liangzhu pottery based on form style recognition. Mech. Mater. **268–270**(2013), 1970–1973 (2012)
3. Maruţoiu, C., Bratu, I., Ţiplic, M.I., et al.: FTIR analysis and 3D restoration of Transylvanian popular pottery from the XVI–XVIII centuries. J. Archaeol. Sci. **19**, 148–154 (2018)
4. Acevedo, D., Laidlaw, D., Joukowsky, M.S.: A virtual environment for archaeological research. In: Proceedings of the 28th Annual International Conference of Computer Applications in Archaeology, April, pp. 313–316 (2000)
5. Gaitatzes, A., Christopoulos, D., Roussou, M.: Reviving the past: cultural heritage meets virtual reality. In: Proceedings of the 2001 Conference on Virtual Reality, Archeology, pp. 103–108 (2002)
6. Zhang, G., Wang, C.: Research of web3D-based ceramic products virtual displaying platform. In: International Conference on Electronic & Mechanical Engineering and Information Technology, vol. 6, pp. 2974–2977. IEEE (2011)
7. Cho, S., Heo, Y., Bang, H., et al.: Turn: a virtual pottery by real spinning wheel. In: International Conference on Computer Graphics and Interactive Techniques, p. 25 (2012)
8. Murugappan, S., Piya, C., Ramani, K., et al.: Handy-potter: rapid exploration of rotationally symmetric shapes through natural hand motions. J. Comput. Inf. Sci. Eng. **13**(2), 1–8 (2013)
9. Vinayak, Ramani, K., Lee, K.J., et al.: zPots: a virtual pottery experience with spatial interactions using the leap motion device. In: CHI 2014 Extended Abstracts on Human Factors in Computing Systems, pp. 371–374 (2014)
10. Lee, J., Han, G., Choi, S.: Haptic pottery modeling using circular sector element method. In: International Conference on Haptics: Perception, pp. 668–674 (2008)
11. Chaudhury, S., Chaudhuri, S.: Volume preserving haptic pottery. In: Haptics Symposium, pp. 129–134. IEEE (2014)
12. Zheng, J.Y., Zhang, Z.L., Abe, N.: Virtual recovery of excavated archaeological finds. In: International Conference on Multimedia Computing and Systems, pp. 348–357. IEEE (1998)
13. Korida, K., Nishino, H., Utsumiya, K.: An interactive 3D interface for a virtual ceramic art work environment. In: International Conference on Virtual Systems and Multimedia, pp. 227–234. IEEE (1997)
14. Han, Y.C., Han, B.: Virtual pottery: an interactive audio-visual installation. In: International Conference on New Interface for Musical Expression, pp. 3–6 (2012)
15. Mustafa, A.W., Ismail, A.F.: 3D virtual pottery environment using hand gesture interaction. In: Proceedings International Conference on Virtual Systems and Multimedia, vol. 3, pp. 1–6 (2018)
16. Vinayak, Ramani, K.: A gesture-free geometric approach for mid-air expression of design intent in 3D virtual pottery. Comput. Aided Des. **69**(C), 11–24 (2015)
17. Park, G., Choi, H., Lee, U., et al.: Virtual figure model crafting with VR HMD and Leap Motion. Imaging Sci. J. **65**(6), 358–370 (2017)
18. Neira, C.C.: Potel: low cost real time virtual pottery maker simulator. In: Virtual Reality International Conference, p. 12. ACM (2016)
19. Huang, H., Zhang, Y.L., Zhang, M., et al.: Research and implementation of digital ceramics museum based VRML. In: International Conference on Computer, Networks and Communication Engineering, pp. 320–322 (2013)
20. Avella, F., Sacco, V., Spatafora, F., et al.: Low cost system for visualization and exhibition of pottery finds in archeological museums. SCIRES-IT **5**(2), 111–128 (2015)
21. Martínez, B., Casas, S., Vidal-Gonzále, M., et al.: TinajAR an edutainment augmented reality mirror for the dissemination and reinterpretation of cultural heritage. Multimodal Technol. Interact. **2**(2), 1–13 (2018)
22. Moore, P.: Learning and teaching in virtual worlds: implications of virtual reality for education. J. Educ. Technol. **11**(2), 91–102 (1995)

23. Han, G., Hwang, J., Choi, S., et al.: AR pottery: experiencing pottery making in the augmented space. In: International Conference on Virtual Reality, pp. 642–650 (2007)

24. Chiang, P.Y., Chang, H.Y., Chang, Y.J.: PotteryGo: a virtual pottery making training system. Comput. Graph. IEEE **38**(2), 74–88 (2018)

25. Wang, T.W.: Empowering art teaching and learning with iPads. Art Educ. **71**(3), 51–55 (2018)

26. Cohen, F., Liu, Z., Ezgi, T.: Virtual reconstruction of archeological vessels using expert priors and intrinsic differential geometry information. Comput. Graph. **37**(1–2), 41–53 (2013)

27. Iqbal, M., Hisham, Q.: Virtual assembly of pottery fragments using moiré surface profile measurements. J. Archaeol. Sci. **32**, 1527–1533 (2005)

28. Kampel, M., Sablatnig, R.: An automated pottery archival and reconstruction system. Comput. Animat. Virtual Words **14**(3), 111–120 (2003)

29. Kumar, G., Sharma, N.K., Bhowmick, P.: Creating wheel-thrown potteries in digital space. In: Huang, F., Wang, R.-C. (eds.) ArtsIT 2009. LNICST, vol. 30, pp. 181–189. Springer, Heidelberg (2010). https://doi.org/10.1007/978-3-642-11577-6_23

30. Gao, Z., Li, J., Wang, H., et al.: DigiClay: an interactive installation for virtual pottery using motion sensing technology. In: International Conference on Virtual Reality, pp. 126–132 (2018)

31. Makransky, G., Terkildsen, T.S., Mayer, R.E.: Adding immersive virtual reality to a science lab simulation causes more presence but less learning. Learn. Instr. **60**, 225–236 (2019)

An Annotation Method for Artwork Attributes Based on Visual Perception

Chenyang Cui[✉]

China Academy of Art, Hangzhou, Zhejiang, China
cuicy@caa.edu.cn

Abstract. In most of the existing online search systems of museums, only the professional knowledge keywords are used to retrieve the artworks. It is a challenge for either professionals or non-professionals. In this paper, we divide the attributes of an artwork into two categories: subjective and objective, and propose a method of attribute annotation for artworks based on the visual perception of the searcher or audiences, which makes the retrieval simpler and more suitable for either professionals or non-professionals.

Keywords: Artworks retrieval · Visual perception · Attributes annotation

1 Introduction

The development of the global economy has promoted the prosperity of the cultural and art market. The Internet has become one of the main ways for people to understand art and watch art exhibitions. At present, artworks search engines generally exist in independent museums or art galleries, such as the Musée du Louvre, the Museum of Modern Art in New York (MOMA), and the National Gallery of Art, China Art Museum, etc. Most of the museums' online search systems are mainly based on some keywords such as the name of the artwork, the author's name, the genre of artwork, and even the serial number in the collections, and sometimes some retrieval systems also provide the user with a selectable list for the corresponding screening search. The search keywords in the collection search interface of the Louvre in France include the artist's name, the serial number of the collection, etc. The keywords combination is carried out in the search interface of the Museum of Modern Art in New York, such a search method requires the searcher to have certain professional knowledge. The method based on the accurate information of the artwork is difficult not only for ordinary users, but even for professionals.

Factually, people can only provide some ambiguous description of the needed artworks, such as a certain artwork to have a certain color, shape, content or material, or similar to an image. Obviously, it is necessary to provide users with an intuitive, accurate search based on visual perception attribute descriptions. Such search applications are currently not available in the existing search engine, which involves the latest research fields such as attribute annotation and attribute learning in the artificial intelligence, machine learning, art psychology, computer vision, etc. To overcome the shortcomings

Z. Pan et al. (Eds.): Transactions on Edutainment XVI, LNCS 11782, pp. 45–54, 2020.
https://doi.org/10.1007/978-3-662-61510-2_5

of many current search engines, constructing a kind of artwork retrieval platform for the general public through the visual perception attributes is very important and urgent, which not only provides users with more accurate and practical information, but also conforms to the "future search". Therefore, in order to enable people to retrieve the wanted artworks from the vast artwork databases, we propose the attribute annotation method based on subjective needs and fuzzy descriptions.

2   Analysis of Artworks

Whether an artist or an ordinary viewer can understand an artwork from the following different perspectives: Why did the artist paint this picture? What kind of cultural background does the painting convey? Is the work realistic? What is the mode of composition, color, strokes of the painting, etc.? For the previous two questions, the viewers need to understand the growth background of the artist, the creative background and the creation theme of the artwork through reading the relevant literature before they can understand the artwork. For the latter questions, the viewers themselves need to have a certain professional understanding of the artwork. Most of time, more audiences have no professional art knowledge, they are used to interpreting the artwork from their subjective feeling.

Figure 1 shows an oil painting named as "The Salisbury Cathedral from the Bishop's Grounds", created in 1820 by the British painter John Constable, 88 × 112 cm, firstly collected in Victoria and Albert Museum in London. This oil painting was completed by Constable on the basis of a sketch he drew in 1811 and was factually commissioned by Bishop Fisher, who was a lifelong close friend and supporter of Constable. The painting depicts Salisbury Cathedral, the tallest Gothic building in the Middle Ages, and also emphasizes the harmonious and equal relationship among the Church, British society and nature. This work, whether in transparency, atmosphere, depth, or the overall effect

Fig. 1. The Salisbury Cathedral from the Bishop's Grounds, John Constable

of the picture, is the classic representative of all Constable's works. The part of the picture is covered by some tall trees, several cows are staying in the shade of the trees and grazing leisurely, the sun shining on the church make the viewer feel spectacular and bright. The vertical towering spires and horizontally extending roofs outline the main lines of the church. In the vast natural environment, it is a perfect expression of the sacredness, tranquility and magnificence of the church.

Figure 2 is "The Church at Auvers", created by the Dutch painter Vincent Van Gogh in 1890. In this painting, Van Gogh made the church, the path, the grass, the sky and the surroundings full of vitality through rough and unrestrained brush strokes. The clouds in the sky were almost frozen, and the ancient church on the ground was distorted and deformed, as if moaning in pain. The painter painted the green ground and the yellow road with contrasting colors, like the waves of the sea. The turbulent ups and downs all reflect Van Gogh's restless and painful heart and the twisted world about to collapse. The painting is rich in color but lacks transparency and light. It is painted with thick outlines and uses dynamic strokes to make the lines smoother. The uneven brush strokes of green, blue, brown, and ochre create an unstable impression, as if the church was about to collapse. The disturbing strokes around the building emphasize the dynamics of the picture, and it seems as if the church is about to be taken away. The seemingly free composition and styling make the viewers feel the passion of the artist at all times. Compared with Fig. 1, Fig. 1 accurately depicts the real reproduction of the church in Salisbury, and Fig. 2 is far from the real church, completely the artist's personal imagination with his unique composition, brushwork, and color. The two works convey two completely different feelings to the audience.

Fig. 2. The Church at Auvers, Vincent van Gogh

Figure 3 is the oil painting "Sunrise" created by the French painter Claude Monet in 1872. The picture is in a thin gray tone. Many details are presented through the seemingly random brushstrokes and inconspicuous colors. The painting depicts the scene of the sunrise harbor covered in morning mist, breaking through the limitations of traditional themes and composition, completely based on the perception of visual experience, focusing only on the color relationship and the influence of external light with the brisk jumping stroke. The painting does not pay attention to the delicate lines, and uses a "scattered" stroke to show the misty scene. The painting uses a tic-tac-toe composition, in which the sun rising slowly and the ships drifting on the sea are at the point of interest of the picture, so that the composition can focus the viewer's attention on the main body of the picture. This work is the most typical of Monet's paintings and is a pioneering work of the Impressionist School.

Fig. 3. Sunrise, Claude Monet

Figure 4 is a panel painting "Francois I" by the Italian painter Titian in 1881. Titian was a representative painter of the Venice school during the late Renaissance period, he inherited and developed the Venetian painting art, and advanced the use of colors, shapes and brush strokes of oil painting to a new stage. Titian's works are bold, majestic, rigorous in composition, rich in colors and bright. Titian's portraits can reveal the inner world of a character. As shown in Fig. 4, the portrait looks very vivid, the brush strokes are active, but the elegance and majesty of the king are fully demonstrated. The highest requirement for a portrait is to convey the character of the person being painted, just as the artists often use exaggerated techniques to create portraits of politicians. Sometimes, portraits are not the image of the person being painted, but more of an expression of the artist's creative method.

As shown in Fig. 5, it is a portrait of a painting-seller Vollard created by Spanish painter Pablo Picasso in 1910. The artist breaks the traditional foreground and background segmentation, and makes the background, space and foreground as a community, splits the face and body of Vollard into the different geometric shapes such as cones, spheres and triangles. In this Picasso's work, although the painted person's faces and costumes are presented by jagged and messy geometric shapes, there is a strong hint of depth, space and volume, the viewer can still feel Vollard's face and distinct personality from the seemingly messy picture. Such creative methods are very amazing.

Fig. 4. Francois I, Titian

Fig. 5. Ambroise Vollard, Pablo Picso

3   Analysis and Annotation of the of Artworks Attributes

Through the analysis of the above works, we can see that when the audience appreciates a work, different people will interpret the same work from different perspectives. As an ordinary audience, he will pay more attention to the subjective feelings provided by the work. For example, is this work cheerful, depressed, serious, or warm? When an artist views the work, he will analyze the work from a professional perspective, such as the composition, color, and strokes of the work. So, this article will divide the attributes of painting works into two categories: objective attributes and subjective attributes. The objective attribute refers to the unique property of the artwork, which is a total of ten items as partially shown in Table 1, including the following: author, nationality, date of birth, title, painting type (oil painting, Chinese painting, sculpture, Illustrator, photography, etc.), materials, genre, size, creation time, picture content. The content of the picture includes figure painting, landscape painting, architectural painting, etc. The character painting is subdivided into portraits, genre painting (social life and customs), historical

Table 1. Attributes of artworks

Attributes of artworks	Objective attributes	Author	
		Birthday	
		Nationality	
		Title	
		Type	Oil painting, panel painting, photography,…………
		Genre	Abstract, realistic,…………
		Material	Canvas, wall,……….
		Size	
		Creation Time	
		Content	Description
	Subjective attributes	Professional	Brush strokes
			Hue
			Color tone
			Texture
			Line
			Composition
			Scene
			……….
		Emotional feeling	Warm	Cold
			Happy	Sad
			Unrestrained	Implicit
			Concise	Cumbersome
			……….	………..

painting, military paintings, religious myths and legends, history, legends. Landscape paintings include natural scenery, urban street buildings, seascapes, etc. These attributes are unique and not affected by the subjective cognition of the viewer.

In addition, we also analyze some subjective attributes of the paintings, as partially shown in Table 1, and construct two sub-categories in the subjective attributes, one is based on the professional subjective attribute description, such as color, and lines of the work, brush strokes, composition, scenes, etc., the other one is the emotional subjective attributes of the general audience, such as works are warm, serious, cheerful and so on. Regarding the subjective professional attributes of the same artworks, different artists will give different evaluations due to differences in their professional backgrounds and different perceptions of art. The evaluation of the subjective emotional feelings of the same work will be different because of the difference in everyone's perception of the art. In order to accurately record the audience's evaluation of the subjective professional attributes of artistic works, In each subjective attribute, two extreme attribute values are set, such as "smooth" and "hurry", "smart" and "clumsy", "extensive" and "fine" in the stroke, "concise" and "complex", "soft" and "rigid" in the line, etc.

This paper constructs one attribute annotation system, which allows any registered member, who is either artist or any ordinary user, to upload his own and the other people's works, and to annotate and edit the attributes of the artworks. The system adopts the method of sliding bars. The attributes of the artwork can be annotated by sliding bars in two extreme attribute values according to the feelings and perceptions of each operator. Any artwork can be annotated by different people, then the attribute set of each artwork is computed by the specific algorithm. Through this system, it is convenient to establish a visual perception attribute set of the original artwork, which will be used for the subsequent semantic-based fuzzy artwork retrieval research. The following three figures show the different interfaces respectively in this system. Figure 6 shows the interface of

Fig. 6. The interface of attribute annotation system

attribute annotation system, in which, Register and Login allows users to own their own account, Upload allows users to upload their own works or the others' works, Annotate can allows users to annotate the attributes of the existing works in the system.

Figure 7 shows the interface of the subjective perception attribute annotation, on the left side of this interface is the artwork which is to be annotated, and on the right side are various visual perception attributes. Through dragging the slider, the corresponding attribute values between the two extreme attribute values of each attribute can be determined, then the user can submit the visual perception attribute of the artwork to the system.

Fig. 7. Visual perception attribute annotation

Figure 8 shows the interface of the objective attribute annotation. The upper part of the interface shows the works that need to be attributed, and the lower part shows the names of various objective attributes. The user only needs to expand each attribute and fill in the corresponding objective attribute content, then submit it to the system.

4   System Implementation

As shown in the introduction of the first part of this article, at present, almost no more attempt is made in the online search system of museums or art galleries to annotate the attributes of artworks. Some sporadic annotations are not enough to form an attribute set of the artwork. Here, we implemented an online attribute annotation system. The process is shown in the following Fig. 9. A registered user can upload his own artworks or the others', then he can annotate any artworks' objective and subjective attributes in the system databases. The environment for the development of attribute annotation system is as following. Web portal server: HP DL360 G5; CPU: Intel 5120 dual-core processor, 1.86 GHz, 1*4 MB L2 Cache, Memory 6 GB; Hard disk: smart array E200i; Sever: Debian Linux, Apache, PHP, MySQL; Client: HTML, Javascript, JQuery.

Fig. 8. Objective attribute annotation

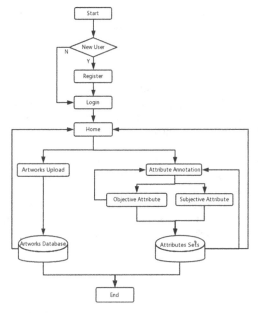

Fig. 9. The process of the attribute annotation

5 Summary

The attribute set of artwork is based on the annotation of a large number of workers. The people's attitude to the annotation, the professional level and the difference in the subjective feeling will affect the validity and quality of the annotated data. Therefore, the attribute annotation method needs to be conducted in further research on how to ensure or improve the validity and quality of the annotated data. Only when the annotation of the attribute is correct and effective, the effective data foundation can be improved and ensure the validity and quality of the annotated data. Only when the data of the attribute set of the artwork is valid, the artwork retrieval based on the description of various attributes can be truly effective.

Acknowledgments. This work is supported by Natural Science Foundations of China (No. 61473256).

References

1. Letts, R.M.: The Renaissance. Cambridge University Press, Cambridge (1981)
2. Woodford, S.: Looking at Pictures. Cambridge University Press, Cambridge (1981)
3. Lambert, R.: The Twenties Century. Cambridge University Press, Cambridge (1981)
4. Reynolds, D.: The Nineteenth Century. Cambridge University Press, Cambridge (1981)
5. Jones, S.: The Eighteenth Century. Cambridge University Press, Cambridge (2012)
6. Shaver-Crandell, A.: The Middle Ages. Cambridge University Press, Cambridge (2012)
7. Mainstone, M., Mainstone, R.: The Seventeenth Century. Cambridge University Press, Cambridge (2012)
8. Woodford, S.: The Art of Greece and Rome. Cambridge University Press, Cambridge (2012)
9. Yang, B., Duanqing, X.: Learning to recognize the art style of paintings using multi-cues. Inf. Technol. Comput. Eng. Manage. Sci. 1, 375–379 (2011)
10. Zhou, C.: Art works Retrieval and Classification. Zhejiang University (2015)
11. Ivanova, K., et al.: Features for art painting classification based on vector quantization of MPEG-7 descriptors. In: Kannan, R., Andres, F. (eds.) ICDEM 2010. LNCS, vol. 6411, pp. 146–153. Springer, Heidelberg (2012). https://doi.org/10.1007/978-3-642-27872-3_22
12. Qi, H., Hughes, S.: A new method for visual stylometry on impressionist paintings. In: Proceedings of IEEE International Conference on Acoustics, Speech and Signal Processing, pp. 2036–2039 (2011)
13. Guo, H., Zheng, K., Fan, X., Yu, H., Wang, S.: Visual attention consistency under image transforms for multi-label image classification. In: Proceedings of IEEE Conference on Computer Vision and Pattern Recognition (2019)
14. Pang, H., Liu, C., Zhao, Z., Zai, G., Li, Z.: Scene image retrieval based on manifold structures of canonical images. Int. J. Pattern Recogn. Artif. Intell. 31(03), 17550005 (2017)
15. Barz, B., Denzler, J.: Hierarchy-based image embeddings for semantic image retrieval. In: Proceedings of IEEE Winter Conference on Applications of Computer Vision (WACV), pp. 638–647 (2019)

Image and Graphics

DPNet: A Dual Path Network for Road Scene Semantic Segmentation

Lu Ye[1,2(✉)], Jiayi Zhu[2], Wujie Zhou[1,2], Ting Duan[2], Sugianto Sugianto[1], George Kofi Agordzo[1], Derrick Yeboah[1], and Mukonde Tonderayi Kevin[1]

[1] School of Information and Electronic Engineering,
Zhejiang University of Science and Technology, Hangzhou 310023, China
yelue@zust.edu.cn
[2] School of Mechanical and Energy Engineering,
Zhejiang University of Science and Technology, Hangzhou 310023, China
948877198@qq.com

Abstract. Road scene segmentation has always been regarded as a pixel-wise task in computer vision studies. In this paper, we introduce a practical and new features fusion structure named "Dual Path Network" for road semantic segmentation. This form aims to reduce the gap between low-level and high-level information, thereby improving features fusion. The Dual Path consists of two subpaths: Context Path and Spatial Path. In the Context Path, we select a pre-trained ResNet-101 model as the backbone and use multi-scale convolution blocks comprise the Spatial Path. Then, we create a fusion residual block and channel attention model to further optimize the network. The results of the experiment confirm a state-of-the-art mean intersection-over-union of 68.5% using the CamVid dataset.

Keywords: Road scene segmentation · Dual path · Fusion residual block · Attention model

1 Introduction

Semantic imaging is one of the most significant research tasks in computer vision. It is used to accomplish the pixel-wise prediction in one image. Fully convolution networks (FCNs) [1] have been popularly applied to improve the spatial resolution in a series of recent seminal studies (e.g., [2–5]), which have obtained promising results for the benchmarks. Encoder-decoder is a classical structure in FCNs, wherein semantic information is first included in the features in the encoding process, following which the decoding process assigns the responsibilities for generating the segmentation results. Typically, the convolution block with pre-trained is used for composing the encoding process that extracts the features, and the resolution is recovered in the decoding process by multiple upsampling operations. However, a shallow network weakens the feature discriminative ability, and thus, we adopt a deep network as the backbone.

Moreover, some problems still need to be overcome. Downsampling operations such as the pooling layer always sacrifice the spatial resolution to ensure the invariance of

© Springer-Verlag GmbH Germany, part of Springer Nature 2020
Z. Pan et al. (Eds.): Transactions on Edutainment XVI, LNCS 11782, pp. 57–69, 2020.
https://doi.org/10.1007/978-3-662-61510-2_6

image transformations. Decreased resolution results in blurred edges and small spurious areas in the output of segmentation. To recover the spatial resolution, many networks use the upsampling or deconvolution path, which can generate high-resolution features for dense prediction. By learning the upsampling process, restoring the low resolution of feature maps to the input resolution of pixel-wise classification is beneficial for accurate boundaries location. By contrast, certain convolution architectures employ dilated convolution, such that the convolution network expands the receptive field exponentially without downsampling. With a muted or even zero downsampling operation, dilated architectures keep the spatial information in the image throughout the whole network. Thus, the architectures serve as discriminative models that classify every pixel on the image. To gain excellent spatial and context features, we introduce a network named Dual Path Network to first extract features separately and then fuse them effectively.

Encoder-decoder architectures, however, lose spatial information during the discriminative encoding, and thus, some of the networks, such as U-Net [6] and FusionNet [7], apply skip architectures to recover the spatial information along the generative decoder path. Empirically, adding more semantic information in low-level features or embedding more spatial information in high-level features can enhance the fusion effect. Therefore, we introduce a new method to reduce the gap between low-level and high-level features, and improve their fusion. The main contributions of this work areas follows:

(1) A novel residual encoder-decoder architecture is proposed for road scene segmentation. It applies a "Dual Path" to extract the spatial features and context features separately.
(2) The later blocks use a new fusion method to fuse the low-level and high-level features effectively, and an attention model is applied to optimize the network.
(3) A pyramid supervision training scheme is proposed. We use the Lovasz function in the branched outputs, and the NLLLoss loss is used in the main output.

2 Related Work

In this section, we briefly review the literature on semantic segmentation. One of the most popular convolutional neural networks (CNNs) is the FCN, and many approaches have achieved high performance for different benchmarks using FCNs. FCNs can outperform many traditional methods by replacing fully connected layers with fully convolutional layers, and concatenating the intermediate score maps in semantic segmentation. Given the structure of FCNs, several works have shown increasing improvements in the results.

To avoid inconsistency within classes, semantic segmentation requires considerable context information to obtain better prediction results. Refs. [5] and [8–10] employ different dilated rates in the convolution layer to capture and fuse different contextual information. Therefore, pyramid structure based on multi-scale has emerged in semantic segmentation.

For example, the atrous spatial pyramid pooling (ASPP) module proposed in [9] used different receptive fields to capture different scales of context information. Ref. [11] improved the neural network using a scale-adaptive convolution layer to obtain adaptive field context information. A discriminative feature network (DFN) adds global pooling on the top of U-shaped architecture to encode the global context [12].

Recently, the attention model has become an increasingly effective method for deep networks as it can use the high-level information to guide the feed-forward network [13–16]. The method in [17, 18] focuses on different scales of information. In this method, similar to the saliency detection network [14], we utilize channel attention to select the features, and then apply it to the recognition task and achieve excellent results. Similar to the DFN [12], it learns the global context as attention and supervises the features.

3 Method

In this section, we first illustrate the structure of our network in Fig. 1, and we use different colors to indicate the various kinds of blocks. Then, we introduce a new form called Dual Path and explain it in detail. Furthermore, we describe the other operations and blocks, such as fusion residual operation, multi-scale supervision and channel attention model (CAM), then explain why they are effective. Lastly, we introduce the loss functions.

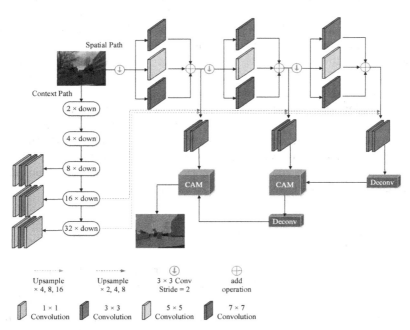

Fig. 1. Overview of DPNet. We choose pre-trained ResNet-101 on ImageNet as the base network of the Context Path ("n × down" refers to one feature with height 1/n and width 1/n of the input image). We use multi-resolution analysis convolution (3 × 3 and 7 × 7 convolutions in parallel to the 5 × 5 convolution) operation to extract features of the Spatial Path. CAM - Channel attention model.

3.1 Dual Path

Our network is illustrated in Fig. 1, which consists of two different features extraction paths. Previous networks such as those in [6] and [7] introduced low-level features with more spatial details, and the high-level features are rich in context information. Based on this characteristic, we divide input images into two paths to extract different feature maps for semantic segmentation, and name them the "Context Path" and the "Spatial Path".

(1) Context Path: Many networks define semantic segmentation as a classification task for each pixel from a microcosmic angle. However, as mentioned previously, similar to the DFN [12], the segmentation can also be viewed as a task that assigns a consistent semantic label to a category of things. It considers segmentation from a macro perspective, and then as the process advances, it proves that the lack of context information is the main factor causing intra-class inconsistency. Therefore, we design a Context Path to extract the context information primarily to prevent this problem. Otherwise, ResNet [19], which is the most frequently cited network in semantic segmentation, shows excellent performance, and the "shortcut" mechanism effectively avoids gradient vanishing and accelerates network convergence. Thus, we use pre-trained ResNet-101 on ImageNet [20] as the backbone of the Context Path to gain additional context information.

(2) Spatial Path: Low-level features, which include concepts such as points, lines, or edges, are significant for predicting the detailed output. We define a Spatial Path as one that does not need many downsampling operations to retain the characteristics of the spatial details. In addition to extracting the spatial information effectively, a large receptive field can extract and generate more abstract features for large objects, while a small receptive field (RF) is suited for small objects. Therefore, we use multi-resolutional analysis [21] that incorporates 3×3 and 7×7 convolutions in parallel to the 5×5 convolution to replace the conventional convolution layer, and choose convolution (stride $= 2$) instead of the pooling operation.

3.2 Reverse Fusion

U-Net [6] as a typical encoder-decoder network that provides state-of-the-art results in the field of medical segmentation. It employs the "long skip connection" on the same position between the encoder and decoder to enhance recovery of the feature maps.

In general, low-level features are too noisy to allow optimization of the fusion process, and thus, a method such as U-Net is not as effective as desired. In contrast, the fusion will be easier and improved if we introduce semantic information into low-level features or spatial information into high-level features, which benefits features fusion. Thus, we obtain the optimization features according to this method.

Using this fusion method, we insert the semantic information into the spatial information and obtain better fusion results. We choose $16\times$ down and $32\times$ down features that are rich in semantic information and adopt upscaling operations with bilinear interpolation to increase the size of the feature maps. Specifically, we set the scale factors as

2, 4, and 8 to expand the features from 16× down to 8× , 4× , and 2× down features. For the same reason, we set the scale features as 4, 8, and 16 to increase the 32× down features to 8× , 4× , and 2× down features.

Then, we use two convolution blocks with shortcut to further optimize the feature maps. There are three input streams and two residual convolution operations for post-processing. The black line indicates the low-level features from the Spatial Path, whereas the blue and orange lines refer to the high-level features from the Context Path. All the input streams are then fused into a whole feature using the "add" operation (i.e., we perform the upsampling operation to increase the size of the two smaller feature maps to the largest resolution of the inputs, and then we use the conv (1 × 1) to reduce the number of channels). Then, the fusion feature is passed sequentially through two convolution blocks, which areas implified version of the convolution unit in the original ResNet.

3.3   Channel Attention Model

We use channel attention to learn the weight distribution, and then apply it to the original features, as shown in Fig. 2. It is noteworthy that each CNN filter acts as a mode detector, and each channel of features in the CNN is activated by the response of the corresponding convolution filter. Each layer of CNNs can output a C × H × W feature map, and different channels of features generate responses to different semantics. Therefore, we add the CAM block after the reverse fusion operation to better focus on the different features.

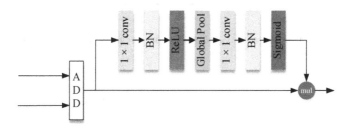

Fig. 2. The CAM block

We use "add" operation to fuse the two inputs and define the fusion features as $I \in W \times H \times C$. In Eq. 1 (we omit the bias operation), convolution operation outputs a feature map $O \in W' \times H' \times C'$, and this set of outputs is defined as $O = [o_1, o_{2,...,}o_{c'}]$, where $P = [p_1, p_{2,...,}p_{c'}]$ is the set of filter kernels, $p_c = [p_{c'}^1, p_{c'}^2, p_{c'}^C]$, $I = [i^1, i^2, \ldots, i^C]$, * refers to the convolution operation, and $p_{c'}^n$ is a 2D spatial kernel representing a single channel of $p_{c'}$ that acts on the corresponding channel of I.

$$O_{c'} = p_{c'} * I = \sum_{n=1}^{C} p_{c'}^n * i^n \tag{1}$$

In order to solve the problem of utilizing channel dependence, we regard the signal to each channel in the output features firstly. Each learning filter uses a local receive field, so each unit of the transformation output O cannot take advantage of contextual information outside that area. This issue increases in severity in the lower layers of the network with smaller receptive field sizes. Therefore, we use global average pooling to generate channel-wise statistics. We generate $g \in C'$ by shrinking O through spatial dimensions $W \times H$. Equation 2 shows the process of calculating the c'-th element of g.

$$g_{c'} = \frac{\sum\limits_{i=1}^{H} \sum\limits_{j=1}^{W} o_{c'}(i, j)}{H \times W} \tag{2}$$

Before the "multiply" operation, we use *Sigmoid* to replace *ReLu* as the activated function, because it can capture the nonlinear interaction between channels flexibly and learn non-mutually exclusive relationships among multiple channels. The relevant expressions are shown as Eqs. 3 and 4. *Sigmoid* can be regarded as a set of ratios $[\lambda_1, \lambda_2, \ldots, \lambda_{c'}]$ to activate the original linear model. Then, we multiply a and the original input feature I to obtain the last output I^*. Figure 3 shows the transformation process (Fig. 4).

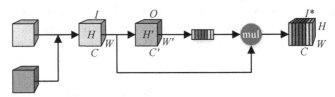

Fig. 3. Transformation process of the CAM blocks (for clear and simple representation, we ignore the second 1 × 1 convolution).

$$a = \sigma(g, p) = \begin{bmatrix} \lambda_1 p_1 \\ \lambda_2 p_2 \\ \vdots \\ \lambda_{c'} p_{c'} \end{bmatrix} \times \begin{bmatrix} g_1 \\ g_2 \\ \vdots \\ g_{c'} \end{bmatrix} \tag{3}$$

$$I^* = a \cdot I \tag{4}$$

3.4 Multi-scale Supervision

The MSS learning over three different scales prevents the gradient from vanishing and optimizes the training parameters. Three intermediate outputs exist from the feature maps of the last three blocks of the Context Path. Each side output score map is computed using an MSS block. We use 1 × 1 convolution to reduce the number of channels to the class number, and all the outputs have different spatial resolutions. They are then fed into a Softmax layer and Lovasz function to predict the output.

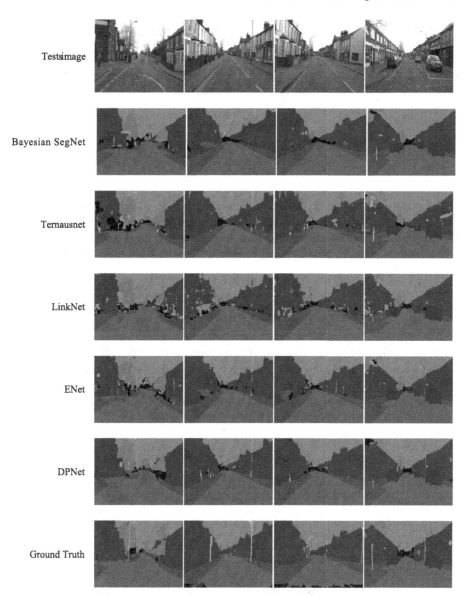

Testsimage

Bayesian SegNet

Ternausnet

LinkNet

ENet

DPNet

Ground Truth

Fig. 4. Qualitative comparison on the CamVid test set

3.5 Loss Function

It is important to choose the proper loss function to ensure prediction results that are as accurate as possible. In semantic segmentation tasks, cross-entropy is the most frequently used function. However, to fit the index better, a new function is proposed, namely the Lovasz-Softmax loss function [22–24]. However, before applying this loss function, the output of the network must be processed through the Softmax function to calculate the probability of each pixel. In our network, we use NLLLoss as our loss function, and its calculation process is the same as that for cross-entropy except for the fact that cross-entropy encapsulates Softmax, but in NLLLoss, we need to add the Softmax layer manually, where N is the number of pixels in the image, y_m is the ground truth class of pixel m, and f is a vector of all network outputs $f_m(i_c)$.

$$f_m(i_c) = \frac{\exp(i_c)}{\sum_{j \in C} \exp(i_j)} \tag{5}$$

$$loss(f) = -\frac{1}{N} \sum_{m=1}^{N} \log f_m(y_m) \tag{6}$$

Direct optimization of the mIoU via Lovasz-Softmax, compare with the traditional cross-entropy, this loss performs better in Jaccard index measurement. According to the derivation process in, they define a segmentation output y^*, ground truth y, and the set of mispredicted pixels of class C as Eq. 7, and the Jaccard Loss can be written as Eq. 8.

$$M_c(y, y^*) = \{y = C, y^* \neq C\} \cup \{y \neq C, y^* = C\} \tag{7}$$

$$\Delta J_c : M_c \in \{0, 1\}^N \mapsto \frac{|M_c|}{\{y = C\} \cup M_c} \tag{8}$$

As per [38], the Lovasz extension of a set function $\Delta : \{0, 1\}^N \to \mathbf{R}$ such that $\Delta(0) = 0$ is defined as Eqs. 9 and 10, where π is a permutation ordering the components of m in decreasing order, that is, $x_{\pi_1} \geq x_{\pi_2} \dots \geq x_{\pi_N}$. $f_m(i_c) \in [0, 1]$ (defined in Eq. 5) is used to construct a vector of pixel errors $m(c)$ for class $c \in C$ (defined as Eq. (11)) and the Jaccard index for class c (see Eq. (12)).

$$\bar{\Delta} : m \in R^N \mapsto \sum_{i=1}^{N} m_i t_i(m) \tag{9}$$

$$t_i(m) = \Delta(\{\pi_1, \pi_2, \dots, \pi_i\}) - \Delta(\{\pi_1, \pi_2, \dots, \pi_i\}) \tag{10}$$

$$m_m(c) = \begin{cases} 1 - f_m(i_c) & if c = y_i, \\ f_m(i_c) & otherwise. \end{cases} \tag{11}$$

$$loss(f(c)) = \overline{\Delta J_c}(m(c)) \tag{12}$$

When considering the class-averaged mIoU metric, we average the class-specific surrogates. Therefore, the Lovasz-Softmax loss is defined as

$$loss(f) = \frac{1}{|C|} \sum_{c \in C} \overline{\Delta J_c}(m(c)) \tag{13}$$

We combine two different loss functions to obtain the best result, as shown in Table 1, and the fourth situation appears to be the most instrumental in back propagation and result convergence. Therefore, we use NLLLoss as the loss function of the final output (which is a full resolution score map) and Lovasz-Softmax as the side outputs (which are downsampled).

Table 1. Results of different combinations

Final output		Side outputs		MIoU
NLLLoss	Lovasz-Softmax	NLLLoss	Lovasz-Softmax	
	√		√	65.2
√		√		66.3
	√	√		66.5
√			√	68.5

4 Experimental Results and Analyses

We evaluate our network on the CamVid dataset. We introduce the dataset first, and then elaborate upon the details of the training parameters to describe our result clearly.

4.1 CamVid Dataset

We train and test on the CamVid dataset. This dataset consists of 701 images of spatial resolution 360 × 480 (height and width, respectively). The images were specified manually. We resize the images into a 352 × 480 resolution and split them into 367 training, 101 validation, and 233 test images. We combine the training and validation subsets for training, and evaluate the semantic segmentation for 11 classes using the test subset.

4.2 Training

The size of each ground truth semantic map is 352 × 480, and we further resize them into three downsampled maps from resolution 44 × 60 to 11 × 15 for multi-scale supervision of the side outputs. The network is trained with the Adam optimizer, and the initial learning rates of all the layers are set to 0.0001, with a decay by a factor of 0.5 in every 50 epochs. We train the model on NVIDIA GeForce GTX 1080T GPU with a batch size of 8 and 500 epochs.

4.3 Analysis Result

We conduct a slicing experiment to demonstrate that all the blocks are effective for our network. The results of different combination modes are shown in Table 2. We retain the Context Path as our baseline and remove other modules in turn. The results demonstrate that the aggregation strategy connects all the blocks, effectively refining high-level and low-level features, without any specifically designed operation. Context features of equal levels are fused by our module, which highly facilitates global support of the query branch.

We also attempt to replace the deconvolution layer with bilinear interpolation, and the results demonstrate that deconvolution provides better performance. Compared with bilinear interpolation, we can better learn useful feature information using the convolution operation in the model, and adapt our model to achieve good results. We compare

Table 2. The results of three main indicators

Model	CA	GA	MIoU
Bayesian SegNet-Basic [25]	70.5	81.6	55.8
DPN [31]	–	–	60.1
ENet [26]	68.3	–	51.3
LinkNet [27]	–	–	55.8
Ternausnet [28]	63.91	86.29	51.27
ReSeg [29]	–	88.7	58.8
DFANet [30]	–	–	59.3
DPNet	74.9	89.0	68.5

our best performance network with other state-of-the-art networks using the same dataset and same task specifications for class-wise accuracy. The comparisons are conducted against Bayesian SegNet-Basic [25], ENet [26], LinkNet [27], Ternausnet [28], ReSeg [29], and DFANet [30].

The results of the three main indicators, namely class accuracy, global accuracy, and MIoU, are shown in Table 3. Our network outperforms the state-of-the-art networks with regard to class accuracy (74.9%), global accuracy (89.0%), and MIoU (68.5%). We introduce the Dual Path form to fuse the different features, unlike previous networks. Simultaneously, feature reuse is possible even as new features are being mined, and the results show that our network not only improves the accuracies, but also consumes less resources. Thus, our result can be considered as excellent. The results for every class are presented in Table 4.

Table 3. The results of slicing experiment. Performances are evaluated by standard meanIoU(%) on CamViddataset.CAM - Channel Attention Model. MSS - Multi-Scale Supervision.

Index	Spatial path	CAM	MSS	Deconvolution	Bilinear interpolation	MIoU
1		√	√	√		66.6
2	√	√	√	√		66.0
3	√		√	√		66.3
4	√	√		√		66.8
5	√	√	√		√	66.2
6	√	√	√	√		68.5

Table 4. The results of 11 classes on CamVid

Model	Sky	Building	Pole	Road	Sidewalk	Tree	Sign symbol	Fence	Car	Pedestrian	Bicyclist
Bayesian SegNet-Basic [25]	91.4	75.1	53	92.5	79.1	68.8	52	44.9	77.7	71.5	69.6
ENet [26]	95.1	74.2	35	95.1	86.7	77.8	51	51.7	82.4	67.2	34.1
Linknet [27]	92.8	88.8	38	96.8	88.4	85.3	41.7	57.8	77.6	57	27.2
Ternausnet [28]	95.2	82.6	27	96.8	81.8	68.2	25.7	66.2	79.8	38.8	72.4
DPNet	96.0	90.1	33.2	97.6	90.6	86.6	52.3	51.6	87.0	64.7	74.2

5 Conclusion

This work proposes a network for road scene segmentation. The method fuses low-level and high-level features using a new form and designs two paths to extract them. It connects a set of convolution layers to effectively refine the high-level and low-level features without a specifically designed operation. We introduce common structures and methods in semantic segmentation, such as context and spatial information, attention model, upsampling, and feature fusion. Furthermore, we show that the reverse fusion and CAM blocks enhance the feature maps and allow us to focus on our goal, respectively. The method also applies multi-scale supervised and different loss functions to further optimize the results. Comparative experiments confirm the state-of-the-art results for our network when using the Camvid dataset.

References

1. Long, J., Shelhamer, E., Darrell, T.: Intelligence, fully convolutional networks for semantic segmentation. In: Proceedings of the IEEE Conference on Computer Vision and Pattern Recognition, pp. 3431–3440 (2015)
2. Chen, L.-C., Papandreou, G., Schroff, F., Adam, H.: Rethinking atrous convolution for semantic image segmentation (2017). arXiv:1706.05587
3. Lin, G., Milan, A., Shen, C., Reid, I.: RefineNet: multi-path refinement networks for high-resolution semantic segmentation. In: Proceedings of the IEEE Conference on Computer Vision and Pattern Recognition, pp. 1925–1934 (2017)
4. Guo, X., et al.: GAN-based virtual-to-real image translation for urban scene semantic segmentation. Neurocomputing. (2019). https://doi.org/10.1016/j.neucom.2019.01.115
5. Yu, H., et al.: Methods and datasets on semantic segmentation: a review. Neurocomputing **304**, 82–103 (2018)
6. Ronneberger, O., Fischer, P., Brox, T.: U-net: convolutional networks for biomedical image segmentation. In: Proceedings of the International Conference on Medical Image Computing and Computer-assisted Intervention, pp. 234–241 (2015)
7. Quan, T.M., Hildebrand, D.G., Jeong, W.K.: Fusionnet: a deep fully residual convolutional neural network for image segmentation in connectomics (2016). arXiv:1612.05360
8. Wang, P., et al.: Understanding convolution for semantic segmentation. In: Proceedings of the IEEE Winter Conference on Applications of Computer Vision, pp. 1451–1460 (2018)

9. Chen, L.C., Papandreou, G., Kokkinos, I., Murphy, K., Yuille, A.L.: DeepLab: semantic image segmentation with deep convolutional nets, atrous convolution, and fully connected CRFs. IEEE Trans. Pattern Anal. Mach. Intell. **40**, 834–848 (2017)

10. Yu, F., Koltun, V.: Multi-scale context aggregation by dilated convolutions (2015). arXiv: 1511.07122

11. Zhang, R., Tang, S., Zhang, Y., Li, J., Yan, S.: Scale-adaptive convolutions for scene parsing. In: Proceedings of the IEEE International Conference on Computer Vision, pp. 2031–2039 (2017)

12. Yu, C., Wang, J., Peng, C., Gao, C., Yu, G., Sang, N.: Learning a discriminative feature network for semantic segmentation. In: Proceedings of the IEEE Conference on Computer Vision and Pattern Recognition, pp. 1857–1866 (2018)

13. Chen, L., Zhang, H., Xiao, J., Nie, L., Shao, J., Liu, W.: SCA-CNN: spatial and channel-wise attention in convolutional networks for image captioning. In: Proceedings of the IEEE Conference on Computer Vision and Pattern Recognition, pp. 5659–5667 (2017)

14. Hu, J., Shen, L., Sun, G.: Squeeze-and-excitation networks. In: Proceedings of the IEEE Conference on Computer Vision and Pattern Recognition, pp. 7132–7141 (2018)

15. Mnih, V., Heess, N., Graves, A.: Recurrent models of visual attention. In: Proceedings of the Advances in Neural Information Processing Systems, pp. 2204–2212 (2014)

16. Wang, F., Jiang, M., Qian, C., Yang, S., Li, C., Zhang, H.: Residual attention network for image classification. In: Proceedings of the IEEE Conference on Computer Vision and Pattern Recognition, pp. 3156–3164 (2017)

17. Chen, L.-C., Yang, Y., Wang, J., Xu, W., Yuille, A.L.: Attention to scale: scale-aware semantic image segmentation. In: Proceedings of the IEEE Conference on Computer Vision and Pattern Recognition, pp. 3640–3649 (2016)

18. Ghiasi, G., Fowlkes, C.C.: Laplacian pyramid reconstruction and refinement for semantic segmentation. In: Proceedings of the European Conference on Computer Vision, pp. 519–534 (2016)

19. He, K., Zhang, X., Ren, S., Sun, J.: Deep residual learning for image recognition. In: Proceedings of the IEEE Conference on Computer Vision and Pattern Recognition, pp. 770–778 (2016)

20. Russakovsky, O., Deng, J., Su, H., Krause, J., Satheesh, S., Ma, S.: Imagenet large scale visual recognition challenge. Int. J. Comput. Vision **115**, 211–252 (2015)

21. Ibtehaz, N., Rahman, M.S.: MultiResUNet: Rethinking the U-Net Architecture for Multimodal Biomedical Image Segmentation (2019). arXiv:1902.04049

22. Berman, M., Triki, A.R., Blaschko, M.B.: The Lovász-Softmax loss: a tractable surrogate for the optimization of the intersection-over-union measure in neural networks. In: Proceedings of the IEEE Conference on Computer Vision and Pattern Recognition, pp. 4413–4421 (2018)

23. Jaccard, P.: Étude comparative de la distribution florale dans une portion des Alpes et des Jura. Bull Soc Vaudoise Sci Nat. **37**, 547–579 (1901)

24. Bach, F.: Learning with submodular functions: A convex optimization perspective. Found. Trends® Mach. Learn. 6, 145–373 (2013)

25. Kendall, A., Badrinarayanan, V., Cipolla, R.: Bayesian SegNet: model uncertainty in deep convolutional encoder-decoder architectures for scene understanding (2015). arXiv:1511. 02680

26. Paszke, A., Chaurasia, A., Kim, S., Culurciello, E.: ENet: a deep neural network architecture for real-time semantic segmentation (2016). arXiv:1606.02147

27. Chaurasia, A., Culurciello, E.: LinkNet: exploiting encoder representations for efficient semantic segmentation. In: Proceedings of the IEEE Visual Communications and Image Processing, pp. 1–4 (2017)

28. Iglovikov, V., Shvets, A.: TernausNet: U-net with VGG11 encoder pre-trained on imagenet for image segmentation (2018). arXiv:1801.05746

29. Visin, F., et al.: ReSeg: a recurrent neural network-based model for semantic segmentation. In: Proceedings of the IEEE Conference on Computer Vision and Pattern Recognition Workshops, pp. 41–48 (2016)
30. Li, H., Xiong, P., Fan, H., Sun, J.: DFANet: deep feature aggregation for real-time semantic segmentation. In: Proceedings of the IEEE Conference on Computer Vision and Pattern Recognition, pp. 9522–9531 (2019)
31. Yu, C., Wang, J., Peng, C., Gao, C., Yu, G., Sang, N.: Learning a discriminative feature network for semantic segmentation (2018). arXiv:1804.09337

Detecting Aging Substation Transformers by Audio Signal with Deep Neural Network

Wei Ye[1]([✉]), Jiasai Sun[1], Min Xu[1], Xuemeng Yang[2], Hongliang Li[2], and Yong Liu[2]

[1] State Grid Zhejiang Electric Power Company's Information Communication Company, Hangzhou, Zhejiang, China
ye_wei@zj.sgcc.com.cn
[2] Institute of Cyber-Systems and Control, Zhejiang University, Hangzhou, Zhejiang, China

Abstract. In order to monitor the aging of transformers and ensure the operational safety in substations, a practical detection system for indoor substation transformers based on the analysis of audio signal is designed, which use computer technology instead of manpower to efficiently monitor the transformers working states in real-time. Our work consists of a small and low cost AI-STBOX and an intelligent AI Cloud Platform. AI-STBOX is installed directionally in each transformer room for continuously collecting, compressing and uploading the transformers audio data. The AI Cloud Platform receives audio data from AI-STBOX, analyses and organizes the data to low-dimensional speech features with STFT and Mel cepstrum analysis. Input the features into a powerful deep neural network, the system can quickly distinguish the working states of each substation transformer before is has serious faults. It can locate aging transformers, command the maintenance platform to quickly release the repair task, thus avoid unforeseeable outages and minimize planned downtimes. The approach has achieved excellent results in the substation aging transformers detection scene.

Keywords: Substation transformer · Speech feature · Deep neural network

1 Introduction

The normal operation of the 10 kV distribution line is of great significance to the power system, which is also an important guarantee to meet the electricity demand of urban and rural residents in China. However, it is undeniable that at present, there are still many problems in operating the 10 kV distribution lines, which seriously affects the operation and management of distribution lines and is not conductive to the development of electric power industry. The 10 kV distribution network is mostly radially arranged and varies in length. There can be many branch units in one line including switching station, high-voltage

© Springer-Verlag GmbH Germany, part of Springer Nature 2020
Z. Pan et al. (Eds.): Transactions on Edutainment XVI, LNCS 11782, pp. 70–82, 2020.
https://doi.org/10.1007/978-3-662-61510-2_7

branch boxes, transformers, low-voltage cabinets, low-voltage branch boxes, etc., reaching dozens or even hundreds of units, and the connections are even more numerous. The 10 kV distribution lines have very complicated paths, and the quality of the equipment is also uneven, so they have different aging rates. Coupled with the influence of external factors, it is difficult to find and deal with the fault quickly in the event.

Transformers are the cornerstone of power generation, transmission, and distribution systems, while transformer aging may brings many adverse effects such as transformer winding burnout, transformer insulation breakdown. Once a fault occurs, it may results in power outage of a specific district or even more serious consequence. Through investigation, we find existing substation monitoring methods rely on manpower. When inspecting the power equipment, electricity workers often judge the states of transformers through experience. The judgment process is actually closely related to the frequency of the sound heard by human. The ability of fault detection is obtained by long-term experience accumulation, that is, some special audio modes correspond to abnormal states. However, it is unrealistic to require patrol personnel to inspect substations frequently as the distance between substations is long and some substations are in remote areas. Such intermittent inspections cannot detect substation abnormalities in real-time and effectively, thus failing to deal with equipment faults in time, resulting in a large risk.

According to the inspection methods of the electricity workers, the aging equipments produce different sounds with normal equipments, which can be an alert of the serious malfunction. The feasibility of audio analysis and the need of real-time inspection inspired us to use a deep learning network to monitor the aging of equipments.

In this work, we adopt an audio classification technology through deep neural network to distinguish transformers states. It is one of the key technologies on how to deal with, analyse and utilize massive audio information. Before classification, any audio signal needs to be extracted its features, which can be classified into time-domain features and spectrum features. Time-domain analysis uses the waveform of the audio signal itself for analysis. Spectral analysis uses the spectral representation of the audio signal for analysis. Common time domain features include Volume Distribution [16], Pitch Contour [16], zero-crossing rate [7], short-time energy [6], short-time autocorrelation function, short-time average amplitude difference, etc. Common frequency domain features include: short-time Fourier transform (STFT) [2], wavelet transform [4], ST [22], etc. There are many feature extraction techniques, including Linear Predictive Analysis (LPC) [8] Linear Prediction Cepstral Coefficient (LPCC) [1], Perceptual Linear Prediction Coefficient (PLP) [10], Mel Frequency Cepstral Coefficient (MFCC) [17], Power Spectrum Analysis (FFT), Relative Spectral Transform of Log Domain Coefficients (RASTA) [11], First Derivative (DELTA).

The development of audio classification technology has a long history, and a series of methods have emerged. In 1977, Sawhney and Maes from MIT Media Lab [21] recorded a dataset from a set of classes including people, voices, subway, traffic and others. Then employed recurrent neural networks and a k-nearest neighbour criterion to model the mapping between the extracted features and categories. Stan Z. Li proposed a new pattern classification method based on a so-called nearest feature line (NFL) in 2000 [15]. Select at least two different samples in each audio category, and make a cepstrum feature of any two samples into a straight line in the feature space, called a feature line. The query input is compared with each feature line separately, and the category of the closest feature line is selected as the query result. Then in 2003, the author proposed a content-based audio classification and retrieval method based on support vector machine (SVM) [9]. Given a feature set consisting of cepstrum features, learn the best class boundaries between classes from the training data by using SVM and finally match by using the distance from the boundary. Recent years, more researchers [3,12–14] used deep network models to classify sounds in the environment. Piczak [18] evaluated the potential of convolutional neural networks in classifying short audio clips of environmental sounds. Roma [19] described a new set of descriptors based on Recurrence Quantification Analysis (RQA), and proposed a framework for environmental sound recognition based on blind segmentation and feature aggregation. These approaches make it possible to classify sound with neural networks for substation transformers.

In this work, a complete online detecting system for indoor substation transformers is introduced, which collects transformers audio data with a non-contact way, extracts its features and judges the current working state of every transformer efficiently by a deep network. It not only saves a lot of unnecessary human resources, but also achieves the purpose of monitoring substation status in real-time.

2   System Structure

Inspired by traditional manual detection methods, we analyse the feasibility of monitoring the transformers by sounds. Besides, sound detection is a non-contact detection method, which can ensure its safety when using in substations. We install AI-STBOX in each transformer room, and the audio data is continuously collected for a certain length of time, compressed and uploaded to the AI Cloud Platform. The AI Cloud Platform analyses and organizes the audio data, extracts the audio features it contains, inputs which into the deep network that we have trained, and finally obtains the results, reflecting the real-time status of each transformer room (aging or not). The whole structure is shown in Fig. 1.

2.1   Introduction of AI-STBOX

AI-STBOX is an audio data acquisition and transmission equipment designed in this work which is shown in Fig. 2. It is installed in transformer rooms for collecting, compressing and transmitting audio data. As we adopt a supercardioid pointing dynamic microphone in AI-STBOX, it directionally captures the sound in front of the microphone and ignores sounds from other directions. Therefore, the noise in the scene is avoided to some extent. AI-STBOX combines a powerful

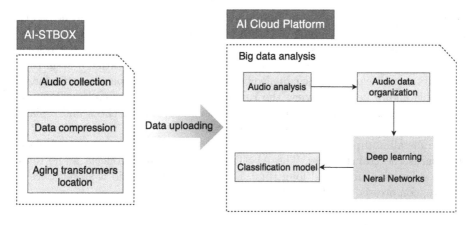

Fig. 1. Introduction of technical solutions. AI-STBOX can upload compressed audio data to AI Cloud Platform, and AI Cloud Platform analyses the data to monitor if the transformer is aging.

1.2 Ghz multi-core processor to make calculations smoother and more responsive. The integrated high-performance DTU module is equipped with 2G/3G/4G network for more reliable data transmission. It is small and low-cost, meet the needs of practical applications in the factory. Moreover, it is industrialized design, dustproof, waterproof and shockproof, and can monitor transformer equipment 24 h a day in real-time.

Fig. 2. AI-STBOX, an audio data acquisition and transmission equipment.

The device is based on an open architecture system design with portability, interoperability, and tailorability. Portability indicates that various computer applications can be ported between various computer systems with open structural features, no matter they are of the same type or model. Interoperability means that the nodes on the computer network can interoperate and share resources, no matter the nodes are of the same type or model. The tailorability shows that the application system running on the low-end machine of the system should be able to run on the high-end machine, and the application system running on the high-end machine can also be run on the low-end machine after being cut. This allows people to view the running status of the low-end machine on the high-end machine. If any AI-STBOX is damaged, we can check and replace it in time to make the whole system more stable.

We install and power up the AI-STBOX in the transformer room and place it towards the transformers. The AI-STBOX continuously collects audio data of the substation in a fixed long time. For example, the audio is stored once a minute, compressed and transmitted to the detection platform.

2.2   AI Cloud Platform

The AI Cloud Platform is used to receive and analyse the raw audio data of each transformer room, and finally identify whether it reflects an aging equipment. The platform mainly includes audio feature extraction, deep network classification, post-processing of detection results.

Audio Feature Extraction. After AI-STBOX transmits the original audio data to the AI Cloud Platform, it is first input to the audio feature extraction module to extract the distinguishing features. The original sound signal is a one-dimensional time domain signal, and it is difficult to visually see the frequency variation pattern. Although the frequency distribution of the signal can be transformed by Fourier transform into the frequency domain, the time domain information is lost so that the change of the frequency distribution with time cannot be seen. In order to solve this problem, many time-frequency analysis methods have emerged. Short-time Fourier transform (STFT) is the most commonly used time-frequency analysis method. It represents the signal characteristics at a certain moment by calculating the signal in time window. The speech signal is time-varying, but at a short time interval, it can be assumed that the signal has hardly changed or changed very little, and the Fourier transform is performed on each time window of the signal to obtain a spectrogram.

To be specific, for the acquired long-time sound data, we seperate the data to several short-time level frames. For each frame, use sliding window to obtain the very short-time interval. For each very short-time interval, performing short-time Fourier transform to get a high-dimensional spectral feature. Short-time Fourier transform is shown as

$$X_a(k) = \sum_{n=0}^{N-1} x(n)e^{-j2\pi k/N}, 0 \leq \mathrm{k} \leq \mathrm{N} \tag{1}$$

where N is the length at Fourier transform, $x(n)$ is the signal to be transformed, X_a is the Fourier transformation result. Figure 3 shows a sample of spectrogram generated by Short-time Fourier transform.

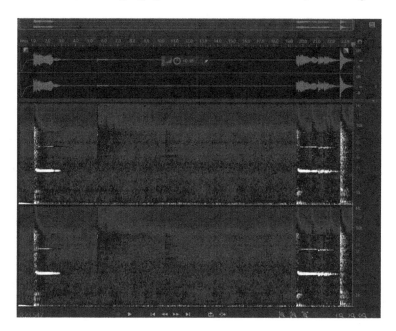

Fig. 3. Audio waveform and spectrogram of an abnormal electric device

As humans' perception of Hertz is not a linear perception, to make it more similar to the human auditory system, the obtained spectrum is usually passed through a Mel filter banks to obtain the Mel spectrum. Several band-pass filters $H_m(k)$ $(0 \leq m \leq M)$, are set, where M is the number of filters. Each filter has a triangular filtering characteristic, and its center frequency is $f(m)$. In the range of Mel frequency, these filters are of equal bandwidth. The transfer function of each band-pass filter is

$$H_m(k) = \begin{cases} 0 & k < f(m-1) \\ \frac{k-f(m-1)}{f(m)-f(m-1)} & f(m-1) \leq k \leq f(m) \\ \frac{f(m+1)-k}{f(m+1)-f(m)} & f(m) \leq k \leq f(m+1) \\ 0 & k > f(m+1) \end{cases} \tag{2}$$

$f(m)$ can be defined as

$$f(m) = \left(\frac{K}{f_s}\right) F_{mel}^{-1}\left(F_{mel}(f_l) + m\frac{F_{mel}(f_h) - F_{mel}(f_l)}{M+1}\right), \tag{3}$$

where f_l is the lowest frequency of the filter frequency range, f_h is the highest frequency of the filter frequency range, f_s is the sampling frequency, F_{mel} function is as

$$F_{mel} = 1125 \ln(1 + f/700). \tag{4}$$

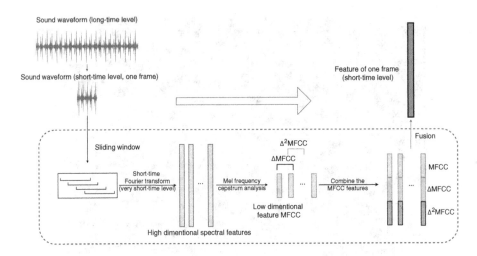

Fig. 4. The process of speech feature extraction. We extract the high-dimensional spectral features of each short-time audio frame and transform them to MFCC, ΔMFCC and Δ^2MFCC. Combine the MFCC features to obtain the final feature.

After the obtained spectrum is passed through the Mel filter bank, perform cepstrum analysis on the Mel spectrum. Specifically, take the logarithm of the powers that we obtained from M Mel filter banks

$$s(m) = \ln\left(\sum_{k=0}^{N-1} |X_a(k)|^2 H_m(k)\right), 0 \leq m \leq M. \tag{5}$$

then perform the inverse Fourier transform, which is generally realized by discrete cosine transform (DTC). Here we use a type of DCT as

$$\acute{C}(n) = \sum_{m=0}^{N-1} s(m)\cos\left(\frac{\pi n(m - 0.5)}{M}\right), n = 1, 2, \ldots, L. \tag{6}$$

In the field of sound processing, Mel frequency cepstrum is a linear transformation of the logarithmic energy spectrum of a non-linear Mel scale based on sound frequency. The difference between cepstrum and Mel frequency cepstrum is that the frequency division of the Mel frequency cepstrum is equally spaced on the Mel scale, which is more similar to the human auditory system than the linearly spaced frequency band used in the normal logarithmic cepstrum. Such a non-linear expression makes it possible to have a better representation of the sound signal in multiple fields.

Use the 2nd to 14th coefficients after DCT as the Mel-frequency cepstral coefficients (MFCC) [20]. The MFCC is the coefficient that constitutes the Mel frequency cepstrum which is widely used for speech recognition. It is proposed by Davis and Mermelstein in the 1980s [5] and has continued to be one of the most advanced technologies. Taking the MFCC composed of 13 coefficients, the difference of the MFCC as ΔMFCC and the difference of the ΔMFCC as Δ^2MFCC, and combine these three parameters to obtain the speech feature representation corresponding to the audio frame. Short-time speech features are composed of a plurality of very short-time MFCC speech feature expressions. Figure 4 shows the process of speech feature extraction.

The MFCC, ΔMFCC and Δ^2MFCC are both of 13-dimensional, and the combination feature corresponding to the very short time is of 39-dimensional. Either compared with the original audio data or the high-dimensional spectral features, the amount of speech feature data corresponding to the very short time is greatly reduced, and thus the calculation amount of the subsequent deep learning classification module is reduced. Then, through a combination of a plurality of very short-time speech features, short-time speech features are obtained.

Deep Neural Network Classification. In this section, the deep neural network is used to learn the mapping relationship between sound features and classification results. As we have shown the possibility to judge the transforms' status by sounds, the deep neural network is a replacement of human to carry out this work. The open-source deep network framework Pytorch is used to build a deep network of 3 fully-connected layers, dropout and batch normalization are also added to improve the generalization ability of the network and accelerate convergence. The structure of the network is shown in Fig. 5.

The pre-acquired and labeled short-time audio features obtained from the substation are used for training, and the label of each short-time audio feature is consistent with the long-time audio to which it belongs. The normal substation sound data is marked as 1, while abnormal sound data is marked as 0. After the input data passes through the audio feature extraction model, the audio features are sent to the deep network in batches, so that the depth model can gradually learn the mapping relationship between the input audio features and the output evaluation. The last layer of the network is connected to the Sigmoid function to ensure the classification result is in range $(0, 1)$. Sigmoid function is as

$$S(t) = \frac{1}{1 + e^{-t}}. \tag{7}$$

The output represents the normal probability corresponding to the input feature. The output value close to 1 indicates that the short-time audio is normal, the working state of the substation tends to be normal. The output value close to 0 means that the short-time audio tends to be abnormal.

Post-processing of Test Results. In order to obtain a more robust result, the output of the deep neural network is encapsulated in a higher level. The deep network output indicates the detection result of relatively short-time audio data. The post-processing part is to combine multiple short-time audio detection results output by the deep networks to obtain the abnormal detection result corresponding to a long-time audio, thereby improving the reliability of the detection method.

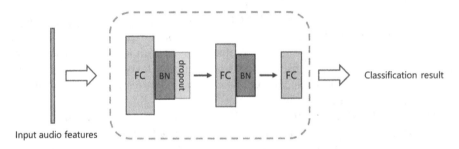

Fig. 5. Deep neural network structure. FC represents the fully connected layer, and BN represents the batch normalization.

Specifically, for each long-time audio we detect the status of multiple short-time audio separately, and when the number of abnormal estimated results is greater than a set threshold, the detection result is determined to be abnormal, and the threshold may make adjustment according to a specific scenario. If the test results for several consecutive long-time intervals indicate that the current transformer is abnormal, it may need to be reported to the relevant maintenance department and check.

Fig. 6. Scene of collecting audio data

3 Experiments

3.1 Experimental Details

AI-STBOXs are directionally installed to collect data with a sampling rate at $f_s = 8\,\mathrm{kHz}$. They provide the on-site audio, which are stored and transmitted to the AI Cloud Platform once a minute. In order to achieve a smoother input, in the process of sampling the original audio data, set the sliding window length of audio frames as 64ms, moves the sliding window every 16 ms. Each sliding window is only 1/4 of the frame length, ensuring consistent feature coverage. The feature input to the network for training is obtained by splicing 50 MFCCs, which means the final input to the network is a feature corresponding to a duration of $(16\,ms*50) - 16\,ms + 64\,ms = 848\,ms$. Finally, the long audio of 1 min can be split into about $60\,s/0.848\,s = 73$ short-time samples. Set $f_l = 300\,Hz$, $f_h = 4\,kHz$, $M = 26$ and $K = 512$ in the Mel filter bank. The deep neural network is built on the open-source framework Pytorch 0.3, and the structure of the network is based on 3 fully connected layers. In order to ensure an efficient training process, the speech features corresponding to all audio frames in the training data are extracted in advance, and the features are directly input to the network.

During the test, 73 short-time frames in one minute are sent to the deep learning network for detection separately. When the number of abnormal states is greater than 5, the state of the current minute can be considered abnormal. When the status of the transformer for 3 consecutive minutes occurs abnormality, it is considered that the maintenance personnel can be notified to check the transformer.

3.2 Dataset

The audio data for experiments are collected in multiple indoor substations for several days. One of the scene of collecting data is shown in Fig. 6. Label the normal or abnormal state of the collected substation data, including the label for each long-time audio. The collected data from aging equipment is abnormal data which marked as 0. The collected data from newly-input substation is normal data which marked as 1. The audio and its corresponding tags are stored for later training of deep learning classification models. A total of about 770 h of data was collected, and about 337w audio samples were generated, of which 80% were used as training data and the remaining 20% were test data. The wav file is an audio file developed by Microsoft Corporation that supports multiple compression algorithms. The audio files used in this experiment were saved in wav format.

3.3 Results

Take 20% of the data as a test set and the rest as training set. The ratio of positive and negative samples is about 7 to 3. The test set includes a total of 158.82 h of data, which is divided into 674,250 short-time samples. The overall test time is 9.13 min. In the 674,250 short-time samples, only 6052 samples are evalued wrongly. The overall accuracy is 99.10%, recall is 99.66%, and precision is 99.10%. The average test of one minute audio data only takes 1.2 ms, the real-time and accuracy can meet the abnormal detection needs of the substation.

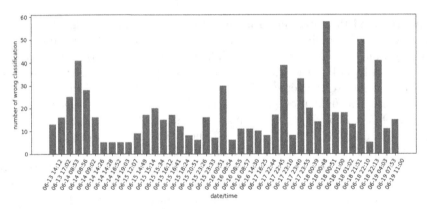

(a) Normal transformer room 1

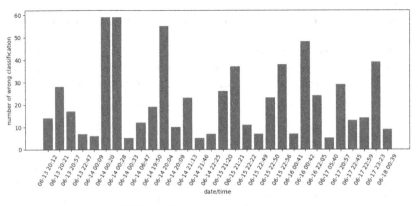

(b) Normal transformer room 2

Fig. 7. Time distribution of the evalued results in normal transformer room 1(a) and 2(b). The histogram shows the number of wrong classification for short-time samples in the corresponding minute.

The estimation results of short-time samples in each minute are counted. When the results of 5 or more than 5 short-time samples are abnormal, the prediction of this minute is judged to be abnormal. We select the examples in which the actual state is normal but judged wrongly. The histogram for two transformer rooms are shown in Fig. 7 (test results for two rooms respectively). It can be seen that the estimation result will not cause an error in several consecutive minutes. We set the threshold time to the maintenance notification is 3 min, which increase the fault tolerance and stability of our system, avoiding frequent overhauls and the waste of human resources.

The experimental results on the test set also show that within the 9238 min data used in the test, only 4 min of abnormal state is divided into normal state. It shows that our method is robust and can be applied in engineering and has practical significance if the samples are sufficient and the amount of data is large enough.

4   Conclusions

In this work, we design a substation transformers online detection system based on the analysis of audio. The system uses a non-contact way to collect transformers sounds and aims to effectively judge aging transformers by sounds through deep neural network, which achieves timely feedback. Through this method, aging equipments can be located, serious accidents can be predicted and avoided to some extent. This makes accident response time greatly shortened, and has significant economic benefits. The results demonstrate the feasibility of applying the proposed method in practical scene.

References

1. Ai, O.C., Hariharan, M., Yaacob, S., Chee, L.S.: Classification of speech dysfluencies with MFCC and LPCC features. Exp. Syst. Appl. **39**(2), 2157–2165 (2012)
2. Allen, J.B., Rabiner, L.R.: A unified approach to short-time fourier analysis and synthesis. Proc. IEEE **65**(11), 1558–1564 (1977)
3. Barchiesi, D., Giannoulis, D., Stowell, D., Plumbley, M.D.: Acoustic scene classification: classifying environments from the sounds they produce. Signal Process. Mag. IEEE **32**, 16–34 (2015). https://doi.org/10.1109/MSP.2014.2326181
4. Daubechies, I.: The wavelet transform, time-frequency localization and signal analysis. IEEE Trans. Inf. Theory **36**(5), 961–1005 (1990)
5. Davis, S., Mermelstein, P.: Comparison of parametric representations for monosyllabic word recognition in continuously spoken sentences (1980)
6. Enqing, D., Guizhong, L., Yatong, Z., Yu, C.: Voice activity detection based on short-time energy and noise spectrum adaptation. In: 6th International Conference on Signal Processing, 2002, vol. 1, pp. 464–467. IEEE (2002)
7. Gouyon, F., Pachet, F., Delerue, O., et al.: On the use of zero-crossing rate for an application of classification of percussive sounds. In: Proceedings of the COST G-6 Conference on Digital Audio Effects (DAFX-00), Verona, Italy, p. 26 (2000)
8. Gray, R.M.: Linear Predictive Coding and the Internet Protocol. Now Publishers, Boston (2010)
9. Guo, G., Li, S.Z.: Content-based audio classification and retrieval by support vector machines. IEEE Trans. Neural Netw. **14**(1), 209–215 (2003)
10. Hermansky, H.: Perceptual linear predictive (PLP) analysis of speech. J. Acoust. Soc. Am. **87**(4), 1738–1752 (1990)
11. Hermansky, H., Morgan, N.: Rasta processing of speech. IEEE Trans. Speech Audio Process. **2**(4), 578–589 (1994)
12. Hinton, G., et al.: Deep neural networks for acoustic modeling in speech recognition: the shared views of four research groups. IEEE Signal Process. Mag. **29**(6), 82–97 (2012). https://doi.org/10.1109/MSP.2012.2205597

13. Kons, Z., Toledo-Ronen, O.: Audio event classification using deep neural networks, pp. 1482–1486, January 2013
14. Lee, H., Pham, P., Largman, Y., Ng, A.Y.: Unsupervised feature learning for audio classification using convolutional deep belief networks. In: Advances in Neural Information Processing Systems, pp. 1096–1104 (2009)
15. Li, S.Z.: Content-based audio classification and retrieval using the nearest feature line method. IEEE Trans. Speech Audio Process. **8**(5), 619–625 (2000)
16. Liu, Z., Wang, Y., Chen, T.: Audio feature extraction and analysis for scene segmentation and classification. J. VLSI Signal Process. Syst. Signal Image Video Technol. **20**(1–2), 61–79 (1998)
17. Logan, B., et al.: Mel frequency cepstral coefficients for music modeling. ISMIR **270**, 1–11 (2000)
18. Piczak, K.J.: Environmental sound classification with convolutional neural networks. In: 2015 IEEE 25th International Workshop on Machine Learning for Signal Processing (MLSP), pp. 1–6, September 2015. https://doi.org/10.1109/MLSP.2015.7324337
19. Roma, G., Herrera, P., Nogueira, W.: Environmental sound recognition using short-time feature aggregation. J. Intell. Inf. Syst. **51**(3), 457–475 (2017). https://doi.org/10.1007/s10844-017-0481-4
20. Sahidullah, M., Saha, G.: Design, analysis and experimental evaluation of block based transformation in mfcc computation for speaker recognition. Speech Commun. **54**(4), 543–565 (2012)
21. Sawhney, N., Maes, P.: Situational awareness from environmental sounds. Project Report for Pattie Maes, pp. 1–7 (1997)
22. Stockwell, R.G., Mansinha, L., Lowe, R.: Localization of the complex spectrum: the S transform. IEEE Trans. Signal Process. **44**(4), 998–1001 (1996)

Statistical Analysis of Principal Dimensions of Dry Bulk Carriers Navigating on Yangtze River Based on Big Data

Jinyu Ren[✉] and Jing Xu[✉]

Naval Architecture and Ocean Engineering, Wuhan Technical College of Communications, No. 6, Baishazhou Road, Hongshan District, Wuhan 430065, Hubei, China
JinyuRen@qq.com, 308810953@qq.com

Abstract. The shipping of the Yangtze River has been developing rapidly, the large scale and standardization of ship dimensions have also developed along with the development of China's economy since entering the 21st century. As a main ship type of the Yangtze River shipping, the dry bulk carriers have changed greatly in the principal dimensions. The characteristics and regression formulas of the principal dimensions in the past are not applicable to the new ship type. On the basis of big data statistics of the dry bulk carrier over 1000 tons navigating on the Yangtze River, the distribution of the transport capacity and distribution of ship age are analyzed, and the linear formula between the gross ton and dead weight is summed up. Then the statistical characteristics of the important parameters of ships such as the ship length, the ship width, the design draft and the diameter of the propeller are analyzed. The research results can provide an important reference for ship design, ship form development, technical standard research and navigation management.

Keywords: Big data · Dry bulk carriers · Principal dimensions · Statistical characteristics

1 Introduction

In 2017, the volume of goods passing through the Yangtze river trunk line reached 2.5 billion tons, among which the dry bulk cargo, mainly metal ore, coal and mining construction materials, accounted for 73%. Dry bulk carrier is undoubtedly the main transport of the Yangtze river. Since the 21st century, the development of economy, the improvement of navigation conditions and the competition of shipping industry have continuously promoted the development of large-scale ships, the research and development of new ship types, and the promotion and application of standard ship types of the Yangtze river. All these factors have led to the new characteristics of the main size of dry bulk cargo ships in the Yangtze river trunk line [1–3]. Existing ship principal dimensions of the distribution is ship design engineers and preliminary design stage to determine the project demonstration of ship principal dimensions of an important reference and the standard technology of ship rules and regulations formulated, navigable

© Springer-Verlag GmbH Germany, part of Springer Nature 2020
Z. Pan et al. (Eds.): Transactions on Edutainment XVI, LNCS 11782, pp. 83–91, 2020.
https://doi.org/10.1007/978-3-662-61510-2_8

channel planning and management has important guiding significance, so it is necessary for the Yangtze river main dry bulk ships on statistical analysis of the distribution of principal dimensions [4, 5].

Based on the big data samples of ships in the Yangtze river trunk line, this paper, after data screening and processing, successively analyzed the distribution of carrying capacity and ship age structure of dry bulk carriers in the Yangtze river trunk line, regression analysis of ship gross tons and deadweight tons, statistical analysis of ship principal scale distribution and statistical characteristics of major scale ratio.

2 Data Processing

By the end of 2017, there were more than 70,000 dry bulk carriers in yingying of the Yangtze river system, from which 12,083 ships of 1,000 tons or more that sailed in the Yangtze river trunk line and were completed in 1998 or later were selected. These data are the research samples of this paper. The acquired samples need to be sorted out and screened to eliminate the samples with errors or defects. The elimination process and methods are as follows [6, 7].

(1) Remove the ship with missing parameters;
(2) Remove vessels with obvious wrong data;
(3) Judge the data according to the correlation between the parameters, and remove abnormal data, such as the square coefficient $Cb > 1$, the load coefficient $Ct > 1$, etc.;
(4) Divide the interval by deadweight ton, and remove abnormal data from the samples within the interval according to the "grubbs" criterion.

After elimination, data samples of 11265 vessels were finally obtained. The following will analyze the distribution of carrying capacity, ship age structure, regression formula between gross tons and deadweight tons, distribution and statistical characteristics of the main scale of vessels and other aspects based on these samples.

3 Transportation Capacity Distribution and Ship Age Structure Distribution

The proportion distribution of the number of ships, deadweight tons and total tons of the sample data is calculated in the deadweight tons interval. The results are shown in Table 1. From the perspective of the number of ships, the number of ships with a deadweight of 1,000 tons to 2,000 tons accounted for 50.3%. With the increase of the range of deadweight tons, the number of ships with more than 15,000 tons accounted for only 1.2%. In terms of the distribution of deadweight ton capacity, 22.2% of ships carrying 1,000 to 2,000 tons of deadweight, evenly distributed between interval 2 and 6, and 6.4% of ships carrying more than 15,000 tons. The proportion distribution of gross tonnage is similar to that of deadweight ton.

Taking 2018 as the base, the age of the ship is obtained by subtracting the year of completion of the ship in 2018. The age structure of the sample data is statistically

Table 1. Distribution statistics of dry bulk carriers in the trunk line of the Yangtze rive

Serial	Deadweight ton range	Number	Proportion	Deadweight tons	Proportion	Gross tons × 10^4	Proportion
1	(1000, 2000]	5662	50.3%	824.4	22.2%	508.4	24.4%
2	(2000, 3000]	2201	19.5%	550.0	14.8%	334.3	16.0%
3	(3000, 5000]	1429	12.7%	563.5	15.2%	327.0	15.7%
4	(5000, 7000]	777	6.9%	462.1	12.5%	259.0	12.4%
5	(7000, 10000]	605	5.4%	509.0	13.7%	270.7	13.0%
6	(10000, 15000]	454	4.0%	560.7	15.1%	271.5	13.0%
7	>15000	137	1.2%	238.7	6.4%	113.6	5.5%
Total		11265	100%	3708.4	100%	2084.5	100%

divided into 4 age intervals, and the results are shown in Table 2. It can be seen that the period from 2008 to 2012 is the peak period of ship construction, with both quantity and deadweight ton capacity accounting for the largest proportion, 41.5% and 44.3% respectively. Between 2013 and 2017, the number of ships built accounted for 22.4%, but the capacity accounted for 31.6%. From the comparison of average deadweight tons, the age of ship [16, 20] was 1888 tons, which gradually increased, and the age of ship [1, 5] was 4,654 tons, 2.5 times of that of [16, 20], reflecting the trend of large ships in the trunk line of the Yangtze river.

Table 2. Distribution statistics of ship age structure of dry bulk carriers along the Yangtze river

Completion date interval	Age range	Number	Proportion	Deadweight tons	Proportion	Gross tons × 10^4	Proportion	Average deadweight ton
[1998, 2002]	[16, 20]	1188	10.5%	224.2	6.0%	141.1	6.8%	1888
[2003, 2007]	[11, 15]	2884	25.6%	667.5	18.0%	408.3	19.6%	2314
[2008, 2012]	[6, 10]	4673	41.5%	1643.8	44.3%	944.1	45.3%	3518
[2013, 2017]	[1, 5]	2520	22.4%	1172.7	31.6%	591.1	28.4%	4654
Total		11265	100%	3708.2	100%	2084.6	100%	

Figure 1 shows the proportion of deadweight tons in different deadweight tons in the four age sections. It can be seen from the figure that the proportion of deadweight tons to 2,000 tons in the age [16, 20] is close to 50%, while the proportion of those over 5,000 tons is only 1%. By comparing the distribution of the four age ranges, it can be found that the proportion of large ships increases with the decrease of the age of ships, among which the ship type with more than 10000 tons did not appear until 2008 and increased sharply after 2013, accounting for 52% of the ship age [1, 5].

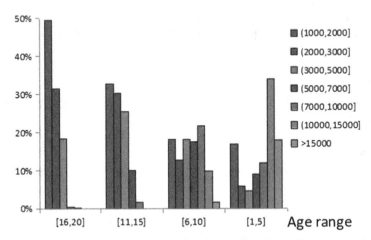

Fig. 1. Transport capacity structure distribution diagram

4   Envelope and Regression Formula of Gross Tons and Deadweight Tons

The gross tonnage of ships is related to the economic evaluation of ships, requirements for equipment allocation of ships, government subsidy policies, etc. It is of great engineering significance to summarize and analyze the relationship between gross tonnage and deadweight ton by using the big data samples of existing ship types. Taking deadweight tons as the abscissa, Fig. 2 gives the scatter distribution of the total tons of dry bulk carriers in the trunk line of the Yangtze river, and makes an approximate upper and lower envelope in the figure. It can be seen from the figure that the lower envelope can be approximately expressed as a straight line. The equation expression of the straight line is given in the figure, while the upper envelope is in the form of a three-fold line. Among them, the upper envelope with a load ton less than 8000 t and the upper envelope with a load ton greater than 13000 t are two segments with different slopes, and the middle broken line between 8000 t and 13000 t is the transition line between the two segments.

Therefore, the regression of the correlation formula between total tons and deadweight tons is divided into two sections with 10000 deadweight tons as the boundary. Table 3 shows the sample number of the two sections, the regression formula and the results of correlation coefficient R2.

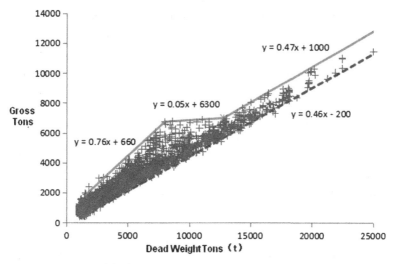

Fig. 2. Gross tons and deadweight tons

Table 3. Regression formula and correlation coefficient of gross tonnage and deadweight ton

Deadweight ton range	Number	Regression formula	Correlation coefficient $R2$
<=10000t	10674	$y = 0.524x + 160$	0.939
>10000t	591	$y = 0.457x + 300$	0.934

5 Statistics of Major Scale Distribution

The sample data of each deadweight ton range are statistically analyzed, and the mean values and distribution ranges of the average deadweight ton, the captain, the width and the design draft of the samples of each range are given. The results are shown in Table 4.

Ratio of principal dimensions is one of the important parameters affecting the performance, the following aspect ratio L/B, wide draught is more important than B/d, Cb square coefficient scale than statistical analysis, and considering the propeller diameter of important influence on the performance of ship propulsion, selection of propeller diameter draft than Dp/d through statistical analysis, in order to provide references for designers. The statistical method is as follows: re-divide the relatively fine deadweight ton interval. The spacing of deadweight ton interval is: 500 tons for ships under 5,000 tons, 1000 tons for ships between 5,000 tons and 10,000 tons, and 2000 tons for ships over 10,000 tons. Then the average value and envelop statistics of the selected parameters for each interval are carried out. The statistical results are shown in Figs. 3, 4, 5, and 6. The x-coordinate value of the curve in the figure is the average deadweight ton value of each interval, and the y-coordinate is the average value of the selected parameters of each interval and the upper and lower envelope value.

Table 4. Main scale distribution statistics of dry bulk carriers along the Yangtze river

Deadweight Ton range		(1000, 2000]	(2000, 3000]	(3000, 5000]	(5000, 7000]	(7000, 10000]	(10000, 15000]	>15000
Average deadweight ton		1456	2499	3943	5947	8413	12350	17423
Captain	verage	57.5	68.8	80.0	90.6	99.5	111.3	127.0
	Scope	46.0–79.5	56.8–88.5	62.2–95.8	78.2–105.6	92.0–114.0	100.8–125.3	120–133.9
Beam	verage	11.3	13.3	14.3	15.8	17.9	20.7	23.0
	Scope	8.8–15.8	11.0–17.2	12.5–17.2	13.8–17.6	15.5–20.8	18.0–23.1	21.6–24.8
Draft	verage	3.4	4.25	5.19	6.07	6.88	7.54	7.86
	Scope	2.00–5.20	2.78–5.90	3.80–6.70	4.35–7.22	5.85–7.65	6.70–8.05	7.55–8.15

It can be seen from Figs. 3, 4, 5, and 6 that the common ground is: the selected parameters of ships below 3,000 tons have a wide distribution range, and with the drastic change of deadweight tons, the distribution of parameters of ships between 3,000 tons and 7,000 tons tends to slow down, while those of ships above 7,000 tons tend to be stable. This is because the scope of the Yangtze river trunk line is wide and the navigable conditions vary greatly, and small ships sail in different navigable waters, resulting in a wide distribution range of the main scale ratio. For large ships, their routes tend to be consistent, mainly in the middle and lower reaches of the Yangtze river trunk line, so the variation of the main scale of ship types tends to be stable.

The variation trend of L/B means is that it first increases with the increase of deadweight tons, then decreases to 5000 deadweight tons, and turns to an increase trend after 13000 tons, in which the lower envelope shows an obvious u-shape between 9000 tons and 17000 tons. The variation trend of the B/d mean value is that it first decreases with the increase of tonnage, and then shows a linear increase trend when it reaches 7000 tons.

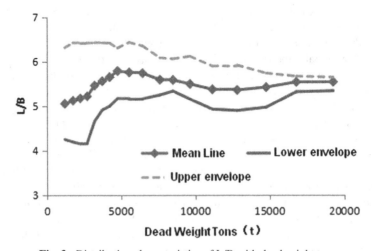

Fig. 3. Distribution characteristics of L/B with deadweight tons

The overall trend of Cb mean value increases with the increase of deadweight tons. The change trend of the Dp/d mean value decreased first with the increase of tonnage, and then showed a stable trend after reaching 6000 tons, and the change was small between 0.34 and 0.36.

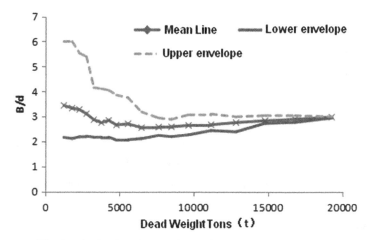

Fig. 4. Distribution characteristics of B/d with deadweight tons

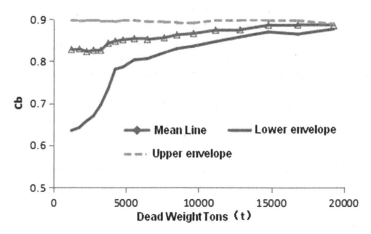

Fig. 5. Distribution characteristics of Cb with deadweight tons

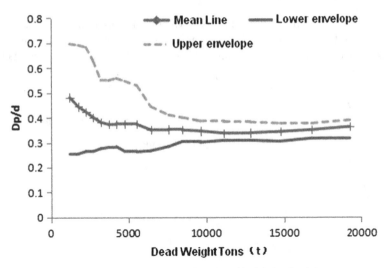

Fig. 6. Distribution characteristics of Dp/d with deadweight tons

6 Conclusion

(1) Through big data statistics of dry bulk carriers in the trunk line of the Yangtze river, the current transport capacity distribution and ship age structure are mastered, which can provide reference for the examination and approval of transport capacity, navigation management and standard formulation;

(2) Based on the statistical regression analysis of the existing big data samples of ship types, a new regression formula for the gross tonnage and deadweight tonnage of dry bulk carriers in the trunk line of the Yangtze river is proposed, which can provide some guidance for ship design, technical standard formulation and management policy formulation.

(3) Based on the length, width, design draft, of ship propeller diameter distribution and the distribution characteristics of some important parameters such as statistical analysis, showing the current Yangtze river main dry bulk ship owners scale survey and statistical characteristics of research and development, the standards for ship design, ship type, navigation management to provide important reference.

References

1. Zhang, L., Meng, Q., Fang Fwa, T.: Big AIS data based spatial-temporal analyses of ship traffic in Singapore port waters. Transp. Res. Part E, 287–304

2. Wang, Q.: Development and design of new ship types for Yangtze river shipping. Ship Sea Eng. **36**(4), 1–4 (2007)

3. Lei, X.: Study on environmental safety assessment of shipping navigation along the Yangtze river. Wuhan: Wuhan University of Technology. (2014)

4. Shuran, Y.: Ecological measures and technical prospects for waterway management of the Yangtze river trunk line. Water Transp. Eng. **1**, 114–118 (2016)

5. Yuan, Y., Feng, Y.: Application of regression analysis method in bulk carrier scheme design stage. Ship Sea Eng. **44**(1), 40–42 (2015)
6. Guan, G., Zheng, M.: Application research of big data technology in the determination of ship's main scale. Appl. Sci. Technol. **45**(2), 1–5 (2018)
7. Wang, C., Yan, L.: Statistical analysis of main scale factors of wind power installation ship. J. Dalian Univ. Technol. **57**(5), 482–487 (2017)

Deep Residual SENet for Foliage Recognition

Wensheng Yan[1(✉)] and Youqing Hua[2]

[1] School of Information Technology Engineering, Taizhou Vocational and Technical College,
Taizhou 318000, China
y1352009@163.com
[2] School of Information Engineering, Jinhua Vocational and Technical College,
Jinhua 321007, China

Abstract. Foliage morphological features are important for plant recognition. However, the foliage shape generally presents big intra-class variations and small inter-class differences. This brings a great challenge to accurate plant foliage recognition. In this paper, we propose a deep residual squeeze-excitation network (R-SENet) for foliage recognition. Firstly, R-SENet learns and obtains the significance levels of each channel of the various convolutional layers in a residual block to recognition tasks via squeeze-excitation strategy. Then, the weights of each channel are rescaled by means of the significances to promote the relevant channels and inhibit non-important channels. Finally, we evaluate the proposed approach on the well-known Flavia dataset for foliage recognition. The experimental results indicate that our approach achieves more accurate average recognition rate (up to 97.86%) and more robustness to noise than other outstanding approaches.

Keywords: Foliage recognition · Residual network · Squeeze-excitation network (SENet) · Rescale weight

1 Introduction

As an important means of plant classification, foliage recognition plays an important role in plant informatics researches [1, 2]. To extract the effective identifying information of plants is a critical link of plant classification and such identifying information includes the local morphological features of plants such as foliage, stem and flower. These features have different values for classification, but compared with other plant organs, foliage features such advantages as easy extraction, high discrimination level and easy conversion to images that can be processed by computer, and is therefore more widely used to perform plant classification tasks [3].

In early researches, the shape of foliages is usually used directly to characterize plants. Oide et al. [4] input the shape features of foliages into a neural network and classified soybean foliages using a Hopfield network and perceptron. Zhai et al. [5] proposed a plant foliage recognition method based on the fractal features of the foliage margin and vein on the basis of fractal theory. Chen et al. [6] extracted the contour-based descriptor, region-based descriptor and mixed descriptor as the shape features

© Springer-Verlag GmbH Germany, part of Springer Nature 2020
Z. Pan et al. (Eds.): Transactions on Edutainment XVI, LNCS 11782, pp. 92–104, 2020.
https://doi.org/10.1007/978-3-662-61510-2_9

of foliages, performed plant species recognition and verified the effectiveness of shape features in plant recognition tasks. As the foliage shape features extracted by these methods are relatively simple, it is difficult to accurately identify the plants with similar foliage shape using these methods. Therefore, Wang et al. [7] proposed a foliage image classification and retrieval method based on chord-features matrices, which could not only achieve accurate characterization of the foliage contour line, but also resist very high noise levels. In addition, texture features having stronger characterization ability have gradually attracted more extensive attentions. Cope et al. [8] extracted the Gabor texture descriptors of foliages and classified 32 plant species using these descriptors. Fu et al. [3] reduced the dimension of LBP features by principal component analysis method, combined LBP features with the statistical shape features of foliages for plant classification and improved the computing speed and foliage recognition accuracy.

Although many remarkable progresses have been made in the researches on plant foliage recognition, there are still many existing issues. During extraction of foliage features, these features are mainly analyzed and identified manually and the issue of difference between different plant species cannot be solved, therefore, there will be deviations when the images of different plants are identified by the same feature. Particularly, under complicated conditions, the recognition accuracy of these methods based on manually processed features will be reduced greatly. By now, major breakthroughs have been made regarding the use of deep learning method in image classification and target recognition and deep learning method can achieve effects much better than those of traditional methods based on manually processed features. As an important structure in deep learning, convolutional neural network (CNN) has gradually become a focus of research. Convolutional neural network can automatically learn and extract image features, improve the generalization ability and reduce the complexity of network structure relying on local receptive fields and weight sharing. In the field of plant foliage recognition, researches on CNN-based methods have just started. Zhang et al. [9] proposed a CNN-based tree foliage recognition method using data augmentation method. Lee et al. [10] and Sugata et al. [11] extracted the features of tree foliages using AlexNet and VGG16 network structures respectively as the backbone networks, which achieved high recognition accuracy on Flavia data set.

Network depth plays a critical role in increasing the accuracy of CNN-based classification methods. These CNN-based foliage recognition methods mentioned above are characterized by simple network structure and small number of layers, which affect the accuracy of plant foliage recognition to a certain extent. However, simply increasing the network depth will cause problems of vanishing gradient and increased training difficulty. Residual network has effectively overcome the difficulties in training deep networks. In order to further improve recognition accuracy and model robustness, this paper proposed a foliage recognition network based on deep squeeze-excitation residual. On the basis of deep residual network and using squeeze-excitation strategy, the level of importance of each channel of the various convolutional layers in a residual block to recognition tasks is determined through calculation and each convolution channel in the residual block is reassigned a weight in such a manner as to promote important channels and inhibit non-important channels. The foliage recognition accuracy is improved significantly.

2   CNN-Based Foliage Recognition Methods

In recent years, CNN-based target recognition and test methods have attracted extensive attention. Compared with traditional methods based on manual processing, CNN-based methods [9–11] have also achieved higher level of recognition accuracy in the field of plant foliage recognition. In this section, these CNN-based methods will be introduced in detail, and the theoretical basis and practical experience for the method proposed in this paper will be provided by collating the network structures and training techniques of these methods.

2.1   Deep-Plant Based Classification Model

In 2015, Lee et al. [10] used CNN-based method for plant foliage classification for the first time and referred to the trained AlexNet network as Deep-Plant classification network. This network consisted of five convolutional layers, three pooling layers, and two fully connected layers at the tail of the network. The detailed Deep-Plant network structure is shown in Fig. 1.

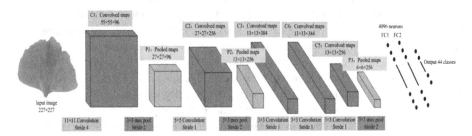

Fig. 1. Deep-Plant network structure

The Deep-Plant model is trained and evaluated on Malaya Kew (MK) data set having 44 plant categories, each image is rotated 45, 90, 135, 180, 225 and 270 degrees respectively, then 2280 images are selected from the data set as training samples and 520 images are selected as test samples. The size of images input to the network is 227×227. In Fig. 1, the blue blocks represent the convolved maps and the yellow blocks represent the pooled maps. $A \times A \times B(C)$ is used to express the configuration of each convolutional layer, where A denotes the size of filter kernel, B denotes the number of output channels and C denotes the convolution stride. The configurations of the five convolutional layers in Deep-Plant model are $11 \times 11 \times 96(4)$, $5 \times 5 \times 256(1)$, $3 \times 3 \times 384(1)$, $3 \times 3 \times 384(1)$ and $3 \times 3 \times 256(1)$ respectively and the configurations of the three pooling layers in the model are $3 \times 3 \times 96(1)$, $3 \times 3 \times 256(1)$ and $3 \times 3 \times 256(1)$ respectively. Receptive fields are related to the size of convolution kernel. The first layer can capture features of larger scale using 11×11 convolution kernel. As the resolution of feature maps will be reduced after pooling, features of large scale can be captured with 5×5 and 3×3 convolution kernels behind the first two rounds of pooling in Deep-Plant model.

Despite its simple network structure, as the first attempt of using CNN for tree foliage classification, Deep-Plant model achieved high recognition accuracy on MK data set and its classification accuracy is 14% higher than that of traditional methods based on manually processed features [12, 13].

2.2 CNN Classification Model with Data Augmentation Function

In order to prevent the model from overfitting, Zhang et al. [9] performed data augmentation of the training images by rotating, translating, scaling, changing the contrast and sharpening, and proposed a CNN classification model based on three-layer convolutional network. This method significantly improved the classification accuracy on Flavia data set, a typical tree foliage recognition data set. In order to verify the effectiveness of data augmentation, the author randomly selected 1585 images from Flavia data set including 32 categories and 1905 tree foliage images for training and the remaining 310 images are used for test. The training samples are augmented by randomly selected data augmentation methods, such as rotating, translating, scaling, changing the contrast and sharpening, and the classification accuracies of the trained model on the original training set, five times augmented, ten times augmented and twenty times augmented training sets are given, as listed in Table 1. It is obvious that the classification accuracy of CNN-based methods can be improved significantly by data augmentation of the training set.

Table 1. Accuracies of classification using training sets of different volumes

Training set	Classification accuracy (%)
Original training set	84.467
6 times augmented training set	88.945
12 times augmented training set	89.856
24 times augmented training set	90.215

In addition to data augmentation during training, reference [9] proposed a CNN network structure based on three-layer convolutions, as shown in Fig. 2. As shown in this figure, the input image is a 256×256 sized color image. In the first convolutional layer, the size of convolution kernel is 3×3, the convolution stride is 1 and there are 32 channels. The first convolutional layer is followed by a max pooling and local response normalization (LRN) layer. The LRN layer can cause the values of strong response to increase, inhibit other neutrons of low feedback level and improve the model's generalization ability. In the second convolutional layer, the size of convolution kernel is 4×4, the convolution stride is 2 and there are 32 channels. The second convolutional layer is also followed by a max pooling and LRN layer. In the third convolutional layer, the size of convolution kernel is 3×3, the convolution stride is 1 and there are 32 channels. It is to be noted that PReLU is used as the activation function behind the first two convolutional layers and the last convolutional layer is directly connected to a Softmax layer for multi-class prediction of foliages.

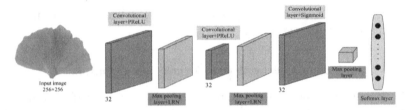

Fig. 2. Structure of three-layer convolutional CNN network

2.3 VGG16-Based Foliage Classification Model

Network depth plays a critical role in the performance of classification models [14]. Based on this consideration, Sugata et al. [11] attempted to use deeper network structures to identity foliages. VGG16 is selected by the author as the backbone network structure, which achieved high-accuracy recognition on Flavia data set (Fig. 3).

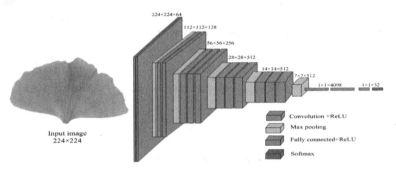

Fig. 3. Foliage recognition based on VGG16 network

The original VGG network [15] is ranked the second in ILSVRC2014, an image classification contest. The structures of many classification networks are improved on the basis of VGG16 thereafter and good effects are achieved. The classification efficiency of VGG networks can be improved significantly by increasing their depths [15]. The original VGG network includes 6 network structures of varying depth, which are expressed as A, A+LRN, B, C, D and E respectively. The network details are shown in Table 2.

These VGG networks mentioned above have similar basic structure but are different in terms of network depth. The detailed structure of each convolutional layer is listed in the table. For example, Conv[256, 3, 1] represents a convolutional layer with 3×3 convolution kernel, convolution stride of 2 and 512 output channels. ReLU is used as the activation function behind each convolutional layer. Three fully connected layers and a softmax layer are connected at the end for target classification. A + LRN network comprises local response normalization (LRN) layer, but tests show that this layer will not increase the classification accuracy and on the contrary, it will increase the computing time. For this reason, the LRN layer is removed from B, C, D and E. In [11], type D VGG structure in Table 1 is used by the author for tree foliage classification and recognition, firstly a pre-trained model is obtained on a large image data set, ImageNet-1K, then the

Table 2. Details of VGG networks of varying depth

A	A + LRN	B	C	D	E
11-layer	11-layer	13-layer	13-layer	16-layer	19-layer
Input (224 × 224 RGB image)					
Conv[64, 3, 1]	Conv[64, 3, 1]	Conv[64, 3, 1]	Conv[64, 3, 1]	Conv[64, 3, 1]	Conv[64, 3, 1]
Max pooling					
Conv[128, 3, 1]	Conv[128, 3, 1]	Conv[128, 3, 1]	Conv[128, 3, 1]	Conv[128, 3, 1]	Conv[128, 3, 1]
Max pooling					
Conv[256, 3, 1] ×2	Conv[256, 3, 1] ×2	Conv[256, 3, 1] ×2	Conv[256, 3, 1] ×2	Conv[256, 3, 1] ×2	Conv[256, 3, 1] ×2
Max pooling					
Conv[512, 3, 1] ×2	Conv[512, 3, 1] ×2	Conv[512, 3, 1] ×2	Conv[512, 3, 1] ×2	Conv[512, 3, 1] ×2	Conv[512, 3, 1] ×2
Max pooling					
Conv[512, 3, 1] ×2	Conv[512, 3, 1] ×2	Conv[512, 3, 1] ×2	Conv[512, 3, 1] ×2	Conv[512, 3, 1] ×2	Conv[512, 3, 1] ×2
Max pooling					
FC − 4096					
FC − 4096					
FC − 1000					
Soft-max					

model is retrained on Flavia data set and the model parameters are adjusted. The results show that this method is apparently superior to traditional methods based on manually processed features and CNN-based methods [9, 10] in terms of classification accuracy, and higher foliage recognition accuracy can be achieved by increasing the network depth.

3 Residual Squeeze-Excitation Network (R-SENet) for Foliage Recognition

In this section, deep squeeze-excitation residual network for foliage recognition will be introduced in detail. The analyses include in the foregoing chapters reveal that it is an effective way to improve classification accuracy by increasing the network depth. However, simply increasing the network depth will cause vanishing gradient problem or exploding gradient problem. Figure 4 shows the accuracies of 20-layer and 56-layer convolutional classifications on CIFAR-10 data set [14] and the size of all convolution kernels used is 3 × 3. From this figure, it can be seen that by simply increasing the network depth, the classification accuracy is not improved and on the contrary, the errors of classification on the training set and test set are increased.

In order to overcome the difficulties with deep network training, He et al. [14] proposed residual networks. If X is defined as the input to a CNN network and the expected output is expressed as $H(x)$, $H(x)$ is deemed as the expected potential mapping and it is difficult for the network to learn this mapping directly. When X is hop transmitted

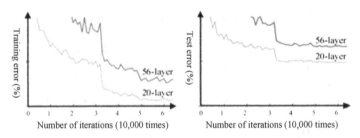

Fig. 4. Classification accuracies of networks of different depths

to the output end as the initial result, the learning target will become $F(X) = H(X)-X$, the difference between the optimal solution $H(X)$ and identity mapping X. The basic block of a residual network is shown in Fig. 5. As the entire network only needs to learn the portion of difference between input and output, residual networks significantly simplify the learning target and reduce learning difficulty.

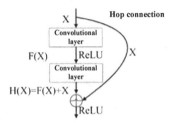

Fig. 5. Basic block of residual network

In residual networks, average pooling layer is used in place of fully connected layer and the number of direct output channels equals the Softmax of the number of target classes, therefore, the computation burden of a 34-layer convolutional residual network is smaller than that of a 19-layer VGG network [14]. Many CNN-based works [16–18] adopt methods other than increasing the network depth.

Unlike these improvements, for the tree foliage recognition network proposed by us, the deep convolutional network is improved from the perspective of the relationship between feature channels, i.e. squeeze-excitation-scale operations [19] are introduced into each residual block in the residual network. Figure 6 shows the detailed structure of the squeeze-excitation-scale block. On the basis, each channel is assigned a weight through scale operation. In simple terms, this block is able to automatically acquire the level of importance of each feature channel by means of learning, promote useful features and inhibit features that are not useful for the current task according to such level of importance.

In Fig. 6, Squeeze, Excitation and Scale denote squeeze operation, excitation operation and weight assignment operation respectively. Firstly, a feature is compressed along the spatial dimension through Squeeze operation and each two-dimensional feature map is compressed into a real number. This real number may be deemed to have global receptive field and the number of output dimensions is equal to the number of feature input

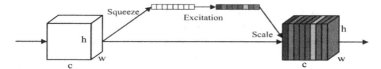

Fig. 6. Squeeze-excitation-scale block

channels. Secondly, a mechanism similar to the gating mechanism in recurrent neural networks is achieved through Excitation operation, the weight of each feature channel is determined through leaning and the correlativity between feature channels is explicitly modeled. The weight output from Excitation operation may be deemed as the level of importance of each channel after feature selection. It is weighted to the previous feature per channel through multiplication and rescaling of the original feature from channel dimension is completed.

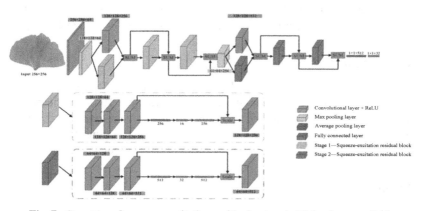

Fig. 7. Structure of squeeze-excitation residual network (Color figure online)

Figure 7 shows the structure of the foliage recognition network proposed by us. The light green block represents Stage 1 squeeze-excitation residual block and the dark green block represents Stage 2 squeeze-excitation residual blocks. Each block consists of three convolutional layers, one average pooling layer and two fully connected layers. In our network, the feature is reassigned a weight before output and input sum operation in each block. If we choose to rescale the feature on the backbone after sum operation, due to the 0–1 rescaling operation on the backbone, vanishing gradient problem tends to occur at the position close to the input layer during optimization of network back propagation, making it difficult to optimize the model.

The size of image input to the network is 256 × 256. The first convolutional layer with 3×3 convolution kernel and 64 channels is used to extract the shallow-layer features and is followed by a max pooling layer to reduce the resolution of feature maps. Stage 1 squeeze-excitation block is connected, a transition layer with 1 × 1 convolution kernel is set up, and the feature map output from this block and the feature map output from stage 1 residual block are input into the next block. Stage 1 squeeze-excitation residual block consisted of three convolutional layers, which had 1, 3 and 3 convolution kernels respectively, the convolution strides are 0, 1 and 1 respectively and the entire block had 256 output channels. Stage 2 squeeze-excitation residual models consisted of three convolutional layers, which had 1, 3 and 3 convolution kernels respectively, the convolution strides are 0, 1 and 1 respectively and the entire block had 512 output channels. The third convolutional layer of each squeeze-excitation block is firstly connected with an average pooling layer, then the weight of each channel is determined through two fully connected layers and sigmoid response, which is eventually weighted per channel to the feature channel of the third convolutional layer of the block through rescaling operation. After passing through the two squeeze-excitation blocks, the output feature maps are sent to the average pooling layer and 512 1×1 feature vectors are obtained. Finally, 32 categories of foliages are classified through fully connected layers and Softmax layer.

4   Experimental Results and Analysis

In this section, the proposed foliage recognition method will be compared with traditional methods based on manually processed features [3–7] and CNN-based methods [9–11].

4.1   Experimental Setting

The model presented in this paper is deployed on Caffe platform [20] and GeForce GTX 1080 GPU with 8 GB memory is used for speed up training. The initial learning rate is 0.001, the weight decay is 0.0005, the batch size is 32, the momentum is set to 0.8, stochastic gradient descent method is used for parameter updating, the size of input image is and the maximum number of iterations is 400,000 times.

4.2   Database

For the test, Flavia, a typical photo library for plant foliage recognition, is selected for algorithm evaluation. This database contains 1905 images of plant foliages of 32 categories. In work [9], it is pointed out that classification accuracy could be improved significantly by performing data augmentation for the training samples. Therefore, data augmentation is performed for Flavia photo library. In this paper, the original training images are augmented by means of 90° rotation, 180° rotation, 270° rotation, horizontal flip, vertical flip, random noise, smoothening and sharpening. From the original Flavia data set, 1405 images are randomly selected as the training set, 220 images are randomly selected as the validation set and the remaining 310 images are used as the test set for evaluation purpose. Figure 8 shows a randomly selected original training image and eight augmented images. The original training image is augmented using several methods that are randomly selected from the eight data augmentation methods mentioned above and then the model is trained using the original training set, 2 times augmented, 4 times augmented and 8 times augmented training sets. The results are shown in Table 3. The results indicate that recognition accuracy can be improved significantly by means of training set data augmentation and this also proves the effectiveness of data augmentation. The results of subsequent tests given in this paper originate from the model trained with 8 times augmented training set.

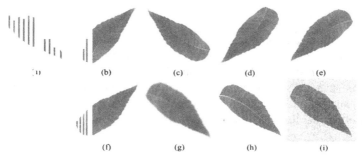

Fig. 8. Samples of augmented training images: (a) original image; (b) 90° rotated image; (c) 180° rotated image; (d) 270° rotated image; (e) vertically flipped image; (f) horizontally flipped image; (g) smoothened image; (h) edge-sharpened image (i) image interfered by Gaussian random noise.

Table 3. Accuracies of classification using training sets of different volumes

Training set	Classification accuracy (%)
Original training set	95.78
2 times augmented training set	97.64
4 times augmented training set	98.65
8 times augmented training set	99.36

The depth of classification network is also of importance to the final classification accuracy. In this section, the accuracies of foliage recognition at varying depth will be given in order to further analyze the relationship between network depth and recognition accuracy. Firstly, assuming that stage 1 consists of K squeeze-excitation blocks and stage 2 consists of G squeeze-excitation blocks, the accuracies of classification at varying network depth are given, as listed in Table 4. When stage 2 consisted of one squeeze-excitation block, i.e. when $G = 1$ and $K = 3$, the classification accuracy reached its peak value of 97.65%. When $K = 3$ and $G = 3$, the classification accuracy reached its peak value of 99.36%. Therefore, both stage 1 and stage 2 of the classification network proposed in this paper consist of three squeeze-excitation blocks.

Table 4. Classification accuracies at different depths

Numbers of K and G values	Classification accuracy (%)
$K = 1, G = 1$	95.55
$K = 2, G = 1$	96.78
$K = 3, G = 3$	**99.36**
$K = 3, G = 4$	99.29

4.3 Results Comparison

In this section, the method proposed herein will be compared with traditional methods [3, 7] and CNN-based methods. The results of comparison are summarized in Table 5. The three CNN-based methods [9–11] are all trained with the same training set. From Table 5, it can be see than CNN-based foliage recognition methods are apparently superior to traditional methods and the recognition accuracy of the method proposed in this paper is remarkably higher than that of other methods.

Table 5. Classification accuracies of different methods

	Methods	Classification accuracy (%)
Traditional methods	[7]	89.69
	[3]	92.75
CNN-based methods	[9]	94.96
	[10]	96.06
	[11]	97.78
Ours		99.36

In addition, the foliage recognition accuracies under different complicated conditions are also compared. The complicated conditions referred to herein mainly include masking and Gaussian noise interference. Firstly, additive white Gaussian noise of 25 variance is added to the images in the test set, the center area of the images is masked at 35% ratio and the classification accuracies of different methods are tested under the conditions of Gaussian noise interference and masking. The results are summarized in Table 6. From Table 6, it can be seen the proposed method is superior to the other methods in terms of classification accuracy.

Table 6. Classification accuracies (%) of different methods under complicated environmental conditions

Type of interference	[5]	[6]	[7]	[3]	[9]	[10]	[11]	Ours
Gaussian noise 25	78.78	82.43	83.09	85.42	90.75	92.08	94.76	97.12
Center masking 35%	83.78	85.11	84.41	86.44	92.13	94.02	95.78	97.56
Average accuracy	82.45	85.09	85.29	87.52	92.19	93.90	95.75	97.86

5   Conclusions

This paper proposed a high-accuracy plant foliage recognition method on the basis of residual neural network and squeeze-excitation strategy. Increasing the depth of classification network is the simplest and most effective way to improve classification accuracy. In order to reduce the difficulty in training deep networks, a deep residual network is used as the backbone network, then the level of importance of each convolutional channel in a residual block to the classification task is determined through learning using the squeeze-excitation strategy, and each channel is reassigned a varying weight. The test results show that the method proposed in this paper is apparently superior to the traditional methods and CNN-based methods in terms of the accuracy of plant foliage recognition.

Acknowledgments. This work was supported by the Major Special Project of Taizhou Vocational and Technical College under Grant No. 2019HGZ02, the Taizhou Science and Technology Project under Grant No. 1902gy31, and the Research Project of Teaching Reform of Zhejiang Province under Grant No. jg20190884.

References

1. Zhang, S., Wang, H., Huang, W.: Two-stage plant species recognition by local means clustering and weighted sparse representation classification. Cluster Comput. **20**(2), 1517–1525 (2017)
2. Lee, S., Chan, C., Mayo, S., Remagnino, P.: How deep learning extracts and learns foliage features for plant classification. Pattern Recogn. **71**, 1–13 (2017)
3. Fu, B., Yang, Z., Zhao, X.L., Shan, Z.L.: Plant leaves recognition method based on dimension reduction local binary pattern and shape features of leaves. Comput. Eng. Appl. **31**(11), 3075–3077 (2011). (Chinese)
4. Oide, M., Ninomiya, S.: Discrimination of soybean foliagelet shape by neural networks with image input. Comput. Electron. Agric. **29**(1), 59–72 (2000)
5. Zhai, C.M., Wang, Q.P., Du, J.X.: Plant leaf recognition method based on fractal dimension feature of outline and venation. Comput. Sci. **41**(2), 170–173 (2014). (Chinese)
6. Chen, L.X., Wang, B.: Comparative study of leaf image recognition algorithm based on shape feature. Comput. Eng. Appl. **53**(9), 17–25 (2017). (Chinese)
7. Wang, B., Chen, L.X., Ye, M.J.: Chord-features matrices: an effective shape descriptor for plant leaf classification and retrieval. Chin. J. Comput. **40**(11), 2559–2574 (2017)
8. Cope, J.S., Remagnino, P., Barman, S., Wilkin, P.: Plant texture classification using gabor co-occurrences. In: Bebis, G., et al. (eds.) ISVC 2010. LNCS, vol. 6454, pp. 669–677. Springer, Heidelberg (2010). https://doi.org/10.1007/978-3-642-17274-8_65
9. Zhang, C., Zhou, P., Li, C., Liu, L.: A convolutional neural network for foliages recognition using data augmentation. In: IEEE International Conference on Computer and Information Technology; Ubiquitous Computing and Communications, pp. 2143–2150 (2015)
10. Lee, S.H., Chan, C.S., Wilkin, P., Remagnino, P.: Deep-plant: plant recognition with convolutional neural networks. In: IEEE International Conference on Image Processing (ICIP), Canada, pp. 452–456 (2015)
11. Sugata, T.L.I., Yang, C.K.: Foliage App: foliage recognition with deep convolutional neural networks. IOP Conf. Ser.: Mater. Sci. Eng. **273**(1), 012004 (2017)

12. Kumar, N., et al.: Leafsnap: a computer vision system for automatic plant species identification. In: Fitzgibbon, A., Lazebnik, S., Perona, P., Sato, Y., Schmid, C. (eds.) ECCV 2012. LNCS, vol. 7573, pp. 502–516. Springer, Heidelberg (2012). https://doi.org/10.1007/978-3-642-33709-3_36

13. Hall, D., Mccool, C., Dayoub, F., Sunderhauf, N., Upcroft, B.: Evaluation of features for foliage classification in challenging conditions. In: IEEE Winter Conference on Applications of Computer Vision, USA, Waikoloa, pp. 797–804 (2015)

14. He, K., Zhang, X., Ren, S., Sun, J.: Deep residual learning for image recognition. In: IEEE Conference on Computer Vision and Pattern Recognition (CVPR), pp. 770–778 (2016)

15. Simonyan, K., Zisserman, A.: Very deep convolutional networks for large-scale image recognition. arXiv preprint arXiv:1409.1556 (2014)

16. Szegedy, C., et al.: Going deeper with convolutions. In: IEEE Conference on Computer Vision and Pattern Recognition (CVPR), USA, Boston, pp. 1–9 (2015)

17. Szegedy, C., Vanhoucke, V., Ioffe, S., Shlens, J., Wojna, Z.: Rethinking the inception architecture for computer vision. In: IEEE Conference on Computer Vision and Pattern Recognition (CVPR), pp. 2818–2826 (2016)

18. Bell, S., Zitnick, C., Bala, K., Girshick, R.: Inside-outside net: detecting objects in context with skip pooling and recurrent neural networks. In: IEEE Conference on Computer Vision and Pattern Recognition (CVPR), pp. 2874–2883 (2016)

19. Hu, J., Shen, L., Sun, G.: Squeeze-and-excitation networks. arXiv preprint arXiv:1709.01507 (2017)

20. Jia, Y., Shelhamer, E., Donahue, J., et al.: Caffe: convolutional architecture for fast feature embedding. In: Proceedings of the 22nd ACM International Conference on Multimedia, USA, Orlando, pp. 675–678 (2014)

Application of Density Clustering Algorithm Based on Greedy Strategy in Hot Spot Mining of Taxi Passengers

Yiping Bao[1], Jianglin Luo[1], and Qingqing Wang[2(✉)]

[1] Jilin Animation Institute, Changchun 130012, China
ice_love713@163.com
[2] College of Optical and Electronical Information,
Changchun University of Science and Technology, Changchun 130000, China
286769533@qq.com

Abstract. In this paper, the greedy strategy is used to improve the density clustering algorithm, which can separate the noise points and deal with the uneven density distribution. In order to further improve the efficiency of density clustering algorithm based on greedy strategy, in this paper, it is applied to mining hot spots of taxi passengers. Firstly, large-scale data are processed, and large-scale data sets are sampled by reservoir, and effective hot data are obtained. Then, the data of 8,000 taxis in an urban area during December 4–8, 2018 are clustered to verify the validity of the proposed algorithm.

Keywords: Density clustering · Greedy strategy · Impounding reservoir sampling

1 Introduction

With the acceleration of urbanization, China's urban scale is expanding. Taxis provide convenience for people to travel, but there will also be taxis empty driving, cruising passengers and other random and inefficient mode of operation, which is a waste of resources for taxis [1]. Clustering analysis is a research branch in the field of data mining. It has been widely used in machine learning, pattern recognition, marketing and other fields. Clustering analysis is a research branch in the field of data mining. It has been widely used in machine learning, pattern recognition, marketing and other fields. According to the characteristics of lightning data, Feng et al. [2] proposed a lightning prediction method based on DBSCAN and polynomial fitting. Wang et al. [3] improved DBSCAN algorithm to mine interest points based on historical trajectory data of mobile smartphone users. Smiti and Eloudi [4] applies the improved DBSCAN algorithm to the domestic taxi dispatching system. Rodriguez and Laio [5] mainly proposes a situation clustering collaborative filtering recommendation algorithm based on undirected graph for mobile terminals, which can detect physical collision of mobile phones by analyzing

Z. Pan et al. (Eds.): Transactions on Edutainment XVI, LNCS 11782, pp. 105–113, 2020.
https://doi.org/10.1007/978-3-662-61510-2_10

users' situation anomaly detection. Du and Zhou [6] applies the multi-dimensional hybrid attribute clustering algorithm based on the mean to the rental system, in order to improve the user's browsing capacity in the search terminal.

Based on the density clustering algorithm, this paper excavates the hot spots of taxi passengers, and then finds and predicts some hot spots of taxi passengers, improves the rate of taxi drivers, and provides decision-making basis for the rational distribution and scheduling of urban taxis.

2 Density Clustering Based on Greedy Strategy

The greedy strategy adaptively finds an appropriate radius for each circle cluster with direct density, and then merges the circle clusters with common points to get the final clustering. The idea of density clustering algorithm based on greedy strategy is to get the global optimal solution through the local optimal solution. Its implementation process is as follows:

Step 1: Get the appropriate search radius and cluster by greedy strategy.
Step 2: Neighborhood queries around clusters;
Step 3: Detection of noise points, so that unreasonable clusters can be queried out;
Step 4: Merge clusters with common points.

2.1 Cluster Discovery and Merging

If each point in the data set can find P min points near its radius R, the steps of cluster discovery and merging are as follows:

In the density clustering algorithm based on greedy strategy, whether the initial point is noise or the core object, the clustering cluster can be obtained, which can ensure the robustness of the algorithm.

For the arbitrarily selected point P, choose on the basis of greedy strategy and find k ($k = P$ min^{-1}) points nearest to P. And then P is the core object of the search. For the elements in set $D = \{d(P, P_1), d(P, P_2), d(P, P_3), \ldots, d(P, P_k), \ldots\}$ sorted from large to small, $d(P, P_k)$ is the Euclidean distance between each point in the set and the core object, and the search radius is $R = d(P, P_k)$.

After obtaining the search radius R, we can get the circle cluster with P as the center and R as the radius. There may be the same points in different clusters, so we need to merge them to adjust the clustering results.

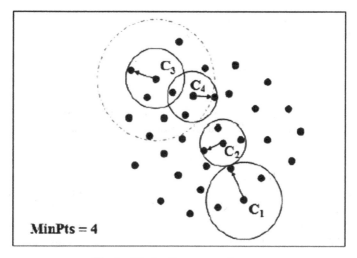

Fig. 1. Cluster discovery and merge

For the above figure, there are four circular clusters, and the radius of each cluster is different. From the above figure, we can see that clusters C_3 and C_4 contain common points, so we need to merge them.

2.2 Neighborhood Query

After merging clusters, all objects in the cluster should be traversed and marked as "clustered". Density clustering algorithm traverses the neighborhood of randomly selected core object P without index. In this case, the efficiency of the algorithm is relatively low. If there are n objects in the data set, the time complexity of density clustering algorithm is $O(n^2)$.

In the density clustering algorithm based on greedy strategy, neighborhood query is used to improve the execution efficiency. For Fig. 1, search for objects not labeled as "clustered" in the neighborhood of cluster d to expand the core objects, find cluster f, and then execute it to guide the query to stop if there is no reasonable and effective direct density within the neighborhood. Neighborhood query can effectively avoid the time-consuming and useless operations caused by blind traversal of data sets. At this time, the time complexity of the algorithm is f, which greatly improves the efficiency of the algorithm.

2.3 Noise Point Recognition

Although the initial point of random selection is an outlier noise point, and the nearest k points can be found to form a cluster, the logical relationship of the cluster is incorrect, which makes the final clustering result unsatisfactory. In this paper, the noise points are detected by relative density, and the average distance d between different points is used as the criterion to measure the relative density.

$$d = \frac{1}{k} \sum_{P_i \in C} d(P_i, P_{k\_nestest(i)}) \tag{1}$$

In formula: P_i is any data point in cluster C, $P_{k\_nestest(i)}$ is the closest k point to P_i, and $d(\cdot)$ is the Euclidean distance. The smaller the value of d, the greater the density around the point.

If the outlier noise point P becomes a nuclear object after certain conditions, by using formula (1), the average distance dis_{avg} between point P and neighborhood midpoint can be calculated. Since the noise points reach the cluster in the direct search density, the edge points in the nearest high-density cluster will be classified into their own clusters. The density of the noise points is relatively low, so the radius R will be much larger than the average distance between the other points in the cluster. That is to say, to satisfy the requirement that $dis_{avg} \ll R$ is the outlier noise point. As shown in Fig. 2, the radius R of point P is much larger than the average distance dis_{avg} between the other points in the cluster, indicating that the point P is an outlier noise point.

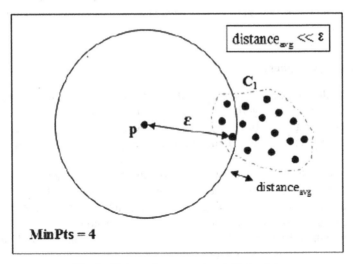

Fig. 2. Noise points as random initial points

3 Application of Density Clustering Algorithm Based on Greedy Strategy in Hot Spot Mining of Taxi Passengers

With the acceleration of urbanization in our country, the mode of transportation in cities is also diversified. As a convenient and flexible way of travel, taxis are welcomed by people. Although the number and scale of taxis are increasing to meet people's travel needs, there are still some problems such as taxi difficulty and low operation efficiency. How to allocate and schedule taxis reasonably has become a research focus to improve taxi operation. The hot data of taxi passengers can accurately describe the trajectory of taxi, so useful information can be found from the data mining technology to solve the problems of taxi no-load, cruise passengers and so on.

Sampling technology can extract appropriate data samples from large data to represent the original data set. There are many sampling methods, such as simple random sampling, reservoir sampling, density deviation sampling and so on. Pool sampling is an efficient sampling method. In view of the unreasonable distribution of taxis and the difficulty of dispatching, this paper uses reservoir sampling to carry out density clustering of greedy strategies, so as to realize the mining of hot spots of taxi passengers.

3.1 Sampling Principle of Reservoir

If the total amount of data is n, k non-repetitive samples are extracted from it, and the probability of each sample being sobbed is $\frac{k}{n}$, then the implementation steps of reservoir sampling are as follows:

Step 1: Construct a reservoir, where the capacity of the reservoir is k, then extract the first i datas from the total data and put them into the reservoir.
Step 2: Starting from the $i + 1$ of the total datas, the probability $\frac{k}{n}$ determines whether the data in the reservoir should be replaced by the data in the reservoir.

$$P_{final\_selected} = P_{now\_selected} \prod_{i=i+1}^{k} P_{not\_replaced} = \frac{k}{n} \tag{2}$$

Among them, $P_{final\_selected}$ denotes the probability that the object will eventually be selected and placed in the reservoir.

$P_{now\_selected}$ denotes the probability that the object appears to be selected.
$P_{not\_replaced}$ denotes the probability that the object is not replaced every time it is replaced.
i is the i ($i \leq k$) data in the original data set, and the probability of being put into the reservoir is 1.

From the $i + 1$th data, the probability that data a may be replaced is $\frac{k}{k+1}$, and the probability of final selection is calculated according to formula (2). When $i = n$, the probability of being extracted is $1 \times \ldots \times \frac{k}{k+1} \times \frac{k+1}{k+2} \times \frac{k+2}{k+3} \times \ldots \times \frac{n-1}{n} = \frac{k}{n}$.
Step 3: All the data in the original data set are traversed, from which kth data are randomly extracted and used as sample data.

3.2 Implementation of Hotspot Mining for Taxi Passenger

The test data selected in this paper are from the GPS driving records of 8000 taxis in a city during December 4–8, 2018. The reason why the historical data is used in this paper is that the real-time GPS data has particularity and is not open to the public. Each taxi will transmit its data to the service desk in 30–70 s and store it in the form of text after acquiring GPS data. Five days are working days. The first four days are mining data and the fifth day is prediction data. This paper verifies the validity of this method. The reason why we did not choose rest days is that the increase of travel on rest days makes the taxi's GPS data change too much, so the general applicability of mining results can not be achieved. The format of GPS raw data storage is shown in the following Table 1.

Table 1. Format of GPS raw data storage

Serial number	Name	Explain
1	Vehicle identification	6-digit representation of vehicle identification
2	Vehicle condition	0 stands for empty car, 1 passenger, 2 fortification, 3 withdrawal, 4 other situations
3	Operation state	0 stands for empty car, 1 passenger, 2 parking, 3 outage, 4 other cases
4	GPS Time	Years, months, days, hours, minutes, seconds, that is yyyyyyMMddHmmss
5	GPS longitude	ddd.dddddddd, Unit is degree
6	GPS latitude	ddd.dddddddd, Unit is degree
7	GPS speed	dddkm/h(size 000-225)
8	GPS azimuth	ddd degree (size 000-360)
9	GPS status	0 represents invalid, 1 represents valid

The uploaded GPS data are 184051, 1, 1, 20171205080137, 116.0955505, 39.9377313, 80, 304, 1. In this paper, according to the characteristics of urban roads, Manhattan distance is used as a distance measurement standard, which is expressed as follows:

Among them, $i(x_i, y_i)$ and $j(x_j, y_j)$ are coordinates of two points in space, representing the linear distance sum of the two points in azimuth. It can be seen from the formula that Manhattan distance is easier to calculate than Euclidean distance, which will improve the calculation efficiency.

After merging the data of GPS traffic records from Dec. 4 to Dec. 7, 2018, the size of the sample is obtained by line statistics. Then the sample is sampled by using reservoir technology. Then the sampled data are clustered by density clustering algorithm based on greedy strategy. The results of clustering are shown in the following Fig. 3.

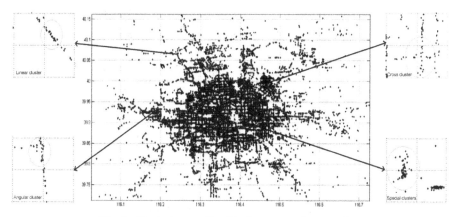

Fig. 3. Clustering results of taxi passenger hot spot data

From the above figure, we can see that the hot spot data of taxis in urban areas are very dense. Because of urban planning, road design and other reasons, the hot spot data of taxis after clustering presents linear, zigzag, cross and other distributions. When clustering, the scattered hot spot data of taxis will be deleted to show the probability comparison of taxi demand among users here. Big, you can guide the taxi driver to this place through the command podium. When finding and guiding taxi passengers in the rush hour, the minimum object parameters will be used to quickly locate the hot spot area of taxi demand, which can better provide potential hot spot area for taxi drivers. During the gentle period of travel, we can use larger object parameters to expand the hot areas of taxi demand, and provide taxi drivers with more accurate positioning of passengers.

3.3 Effectiveness of Predicting Passenger Hot Spot Areas

After mining and clustering the historical data, the early peak passenger hot spot data on December 8, 2018 is forecasted to illustrate the effectiveness of this algorithm in forecasting. This paper takes the urban main road as an example to predict, and the clustering results of historical passenger hotspot data are shown in Fig. 4.

Figure 4 is a clustering of hot spot data of taxis in early peak, so the selected parameters are the smallest object. After removing the scattered noise points, 30 clusters are obtained, that is, hot spot area of passengers. In the process of hitting rate analysis of passenger hotspot areas, the hotspot areas formed by historical data are represented by grey shadows, as shown in Fig. 5. The clustering results of the early peak on 8 December 2018 are obtained.

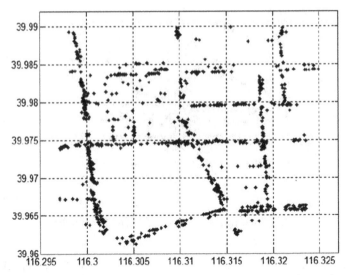

Fig. 4. Clustering results of historical passenger hotspot data on urban main roads

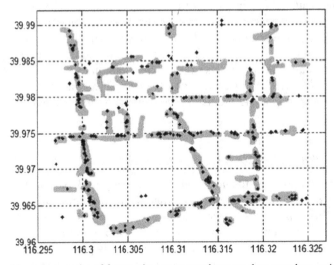

Fig. 5. Clustering results of forecasting passenger hot spot data on urban main roads

Figure 5 shows that the predicted clustering hot spot area is basically consistent with the actual area, only a small amount of data is outside the cluster, which illustrates the effectiveness of the proposed algorithm.

The coverage hit rate is 83.51% by clustering all taxi hotspot data and statistical prediction results. Changkai et al. [7] proposed a taxi hotspot area discovery method based on functional area division. The coverage hit rate was 73.98%. From this, we can see that on the basis of reservoir sampling, the density clustering algorithm based on greedy strategy can better mine the hot spots of taxis, and maintain a high hit rate in finding and predicting passenger demand. It can provide a good basis for reasonable distribution and real-time scheduling of taxis, and solve the problem of taxi difficulty for users.

4   Conclusion

In view of the limitation of density clustering algorithm, this paper uses greedy strategy to improve it. Through this improved algorithm, we can randomly search for core objects and query their neighborhoods. The density clustering algorithm based on greedy strategy is applied to hot spot mining of taxi passengers. In order to improve the clustering efficiency of large-scale data, this paper uses the reservoir sampling technology to sample data sets and get effective hot data. Then the density clustering algorithm based on greedy strategy is used to mine the data of taxi passenger hotspot areas. The final test results show that the algorithm is effective and has high hit rate in finding and predicting taxi hotspot areas.

References

1. Jahirabadkar, S., Kulkarni, P.: Algorithm to determine ε-distance parameter in density based clustering. Expert Syst. Appl. **41**(6), 2939–2946 (2014)
2. Feng, W., Zhu, Y., Guo, J., et al.: Short-term lightning prediction based on improved DBSCAN method and polynomial fitting. Comput. Eng. Sci. **36**(10), 2028–2033 (2014)
3. Wang, Z., Hannah, Song, H., et al.: Mobile user interest point extraction method based on improved DBSCAN. J. Xi'an Univ. Posts Telecommun. **20**(6), 102–105 (2015)
4. Smiti, A., Eloudi, Z.: Soft DBSCAN: improving DBSCAN clustering method using fuzzy set theory. In: International Conference on Human System Interaction, pp. 380–385. IEEE (2013)
5. Rodriguez, A., Laio, A.: Clustering by fast search and find of density peaks. Science **344**(6191), 1492 (2014)
6. Du, D., Zhou, F.: Hybrid collaborative filtering algorithm based on TimeRBM and item attribute clustering. Comput. Appl. Res. **2**, 22–26 (2018)
7. Changkai, Wang, A.: Research on taxi hotspot area discovery method based on functional area division. Comput. Knowl. Technol. (25), 5571–5575 (2013)

VR/AR

A Virtual Marine Ranch for Underwater Tourism

Jiahui Liu[1], Jinxin Kang[1], Pengcheng Fu[2], and Hong Yan[2(✉)]

[1] School of Art and Design, HaiNan University, Haikou 570228, China
774301240@qq.com, 523727340@qq.com
[2] State Key Laboratory of Marine Resource Utilization in South China Sea,
HaiNan University, Haikou 570228, China
{pcfu,yanhong}@hainanu.edu.cn

Abstract. This research mainly explores the visualization of marine ranch in the tourism industry and provides a three-dimensional, dynamic, and visual model for the tourism development of coastal cities. The behavioral characteristics and environment of fish swarm are simulated in virtual marine ranch to shorten the distance between users and marine ecology. This research uses CINEMA 4D and Unity3D game engines to establish virtual scenes of marine fisheries and characteristic fisheries. The interaction between the user and the virtual scene is realized through C# programming and HTC VIVE head display device. Users can visually observe changes in fish swarm and surrounding environment.

Keywords: Virtual reality · Marine ranch · Environmental simulation · Fish swarm simulation

1 Introduction

Wuzhizhou is a particular geographical island in China, which is suitable for both the development of fisheries and tourism [1]. The marine ranch [2] can combine fishery and tourism to form marine ranch tourism [3] (As shown in Fig. 1, Marine Ranch Tourism on Wuzhizhou Island, Hainan, China).

Nowadays, with the development of information technology, tourism informatization has also received more and more attention from the world, and interactive tourism experiences frequently appear in the public view. For example, in January 2016, Thorpe Park and Alton Towers in the UK announced the launch of a virtual reality experience of a roller coaster and ghost train. In March 2016, Yilong, an online travel service provider, released hotel VR experience videos, introducing VR technology in user experience, and providing users with an "unseen prophet, immersive" service when choosing a hotel. Researchers have found that virtual reality travel experiences make content more exciting and more comfortable for tourists to explore more fun to travel.

Besides, the virtual simulation experience enhances the sense of substitution for visitors, allows visitors to close contact with things they have never seen before. In early 2018, the London Museum of Natural History and Sky VR studio launched the "Hold the World VR" experience. Visitors can use the VR headset and controller to

© Springer-Verlag GmbH Germany, part of Springer Nature 2020
Z. Pan et al. (Eds.): Transactions on Edutainment XVI, LNCS 11782, pp. 117–125, 2020.
https://doi.org/10.1007/978-3-662-61510-2_11

visit the conservation center, herbarium and earth science library, and access to the rare and priceless Exhibits. Virtual reality can be combined with many high-risk industries to create a lively, convenient and safe environment for users. Yang Liu set up a virtual ship interactive experiment platform in 2018. Through the four-in-one teaching model of virtual simulation experiments, flipped classroom teaching, shipyard internship, student scientific and technological innovation, the problems of lack of innovation in high-risk teaching methods, low learning interest of students, and lack of high-quality teaching resources were solved [4].

Fig. 1. Marine ranch of Wuzhizhou Island

2 Overview

Based on the above discussion, this study takes the virtual simulation of the marine ranch as the starting point, designs the environment and fish behavior of marine ranch. The aim is to simulate marine pastures and create an excellent interactive environment as an "agency." Users can have an immersive experience. Besides, this research can enable users to obtain an intuitive marine experience in a safe environment, allow users to experience the unknown factors in the ocean fully, and obtain a better travel experience and vivid marine knowledge.

The virtual simulation of the marine ranch should be based on the real situation of Wuzhizhou Island. In order to establish a more realistic regional virtual marine ranch. During the experiment of setting up a virtual marine ranch, we consulted fish models (Fig. 2), such as Great Blue Shark, Horse Mackerel, Grouper, Angelfish, Cobia, et al. [5]. According to these prototypes, we used C4d to design and build a three-dimensional dynamic fish model.

In the environmental simulation design, we mainly considered two aspects of solid facilities and fish breeding in the water. We mainly modeled artificial fish reefs and underwater plants of different shapes. In terms of the user's mobile mode, the click position designed at the initial stage is changed to grab mobile after testing, which is more in line with the real posture of human swimming in the sea, combined with HTC VIVE head display equipment and subtle sound in the water to make it more real [6]. In the experiment, we tried to demonstrate and simulate the marine life from birth to death, from hunger to satiety, and also encountered problems such as model piercing and increasing physical volume. The simulation of marine life is realized by the activities related to marine life, shown in Fig. 3.

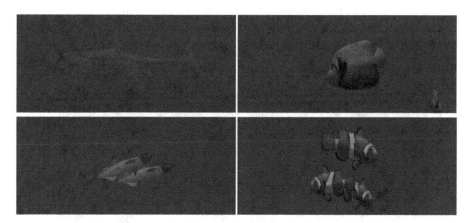

Fig. 2. Fish models

3   Fish Simulation of Marine Ranch

Fish simulation is the most important part of virtual ocean tourism. Since fish swarm is a self-organizing group, there is no fixed leader. The observation and research results of self-organized groups, including birds, insects, and fish swarm, show that individuals in the group may have an orderly and coordinated movement state under the conditions of simple behavior rules [7]. For example, in order to avoid the collision, individuals must keep a minimum distance, attract or repel each other in the group [8]. The influence and importance of these natural factors in group behavior are worth exploring and researching.

For this reason, scientists have carried out a lot of experiments and observations, photographed and recorded the mutual positions of fish groups in the process of long-term swimming. For example, the British Aberdeen Institute of Oceanography built a circular flume to continuously observe Scomber japonicus swimming for a long time to understand the population structure [9]. According to Simon Hubbard, schools of fish are interactive and self-organizing particles, and individual fish are governed by two forces [10]. Breder defines the fish as a specific state of motion in which each fish moves in the same direction and at a uniform speed. The factors of individual spacing in the fish group were: distance was more significant than the critical value and showing attraction; the distance was less than the critical value and showing repulsion [11]. Viscido studied the effects of fish numbers on fish behavior and individual fish interactions [12]. Besides, some researchers have improved the artificial fish swarm algorithm from different aspects. Park using an adaptive parameter adjustment method to improve the algorithm and proposed an improved artificial fish swarm algorithm [13]. Zhang adopts the optimal individual retention strategy to improve foraging behavior, and at the same time, improves the clustering behavior and rear-chasing behavior in the algorithm, and proposes an improved artificial fish swarm algorithm [14]. These improved algorithms improve the effect of the basic artificial fish swarm algorithm to a certain extent, but in the marine ranch environment, the user factor in the existing artificial fish swarm algorithm has not been well solved. Therefore, there is a lack of interaction between fish and users, which makes the user unable to achieve a deeper level of the tourism experience.

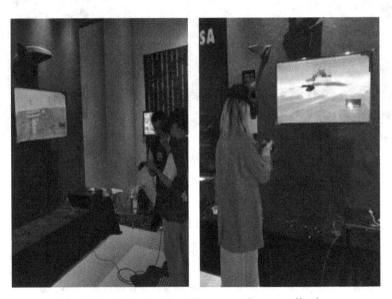

Fig. 3. Part of user's interactive process in our application.

In this study, we will use the concept of the IBMs (Individual Based Models), in which fish will naturally gather in groups during swimming, which is also a living habit formed to ensure the survival of groups and avoid hazards. The formation of a school of fish is also a vivid example. Reyno lds believes that the formation of a cluster of birds and fish does not require a leader. It only needs to follow some local rules of interaction for each bird or fish, and then the cluster phenomenon as a whole pattern from the individual local mutual Emerging in action. There are three rules adopted by Reyno LDS:

Separation Rule: Try to avoid overcrowding with nearby partners;

Alignment Rule: Try to be consistent with the average direction of neighboring partners;

Cohesion Rule: Try to move towards the center of the nearest partner.

We treat the individuals in the group as independent units, follow the exclusion-parallel-attraction rule in the Reynolds model and consider collision avoidance as the first prerequisite, including avoiding collisions between fish swarm and users, shown in Fig. 4.

Fig. 4. Marine fish swarm simulation

Natural Attraction User U to Fish F

When the fish swarm is in a state of non-frightening, the fish begins to surround the user, and the radius around it is $r_d = \text{dis}(u, f_i)$; u represents the current coordinates of the user, and f_i represents the coordinates of the i-th fish.

(1) The radius of the fish around the user gradually increases with time. After t time, the radius around the user is $r_u = r_d - \alpha * u$, and α is the weight coefficient;

(2) When the user and the fish reach a strong repulsive distance, the forward direction of the fish becomes mirror refraction, that is, the fish becomes horrified, and the duration is Δt.

Frightening Rules for Fish Swarm

When the distance between the user and the fish is less than the strong repulsive distance, that is, $r_s > r_u$, the duration of the panic state is $\Delta t' = (1 + \beta * (r_s - r_u) * \Delta t$, and β is the weight coefficient.

Through the improvement of the fish swarm algorithm, the fish swarm is proactively approaching the user, so that users are immersed in it and get a better experience.

4 Environmental Simulation of Marine Ranch

The design of the marine ranch environment mainly considers the following two aspects [15]: solid facilities installed in the sea and inhabiting or breeding of fish. Marine ranch provides necessary and safe habitats for aquatic organisms such as fish to gather, bait, breed, grow and avoid enemies, to protect fishery resources and increase catches.

In the design of solid facilities installed in water, we mainly target artificial fish reefs of different shapes, such as Concrete Reef (mainly concrete, strong plasticity, very common), Stone Reef (pre-processed natural stones into strips of stone and then built into the required type of reef, such as the stone reef in Guangdong Province), small Steel Shipwreck Reef (the frame reef made of steel material is convenient for transportation and production) [16], we did some simple analysis on them. Then we designed and planted Sargassum, Eucheuma, Gracilaria, and other algae around there, to simulate the appropriate type of artificial reef, shown in Fig. 5.

In addition, the research on the ambient lighting model has been a hot spot in the simulation of the marine environment, and it is also one of the main problems in the construction of various scenes of the virtual simulation system. The establishment of lighting models in the ocean directly affects the "immersion" in the VR experience, and also affects the presentation of marine life model details. First, the light illumination model is divided into local illumination and overall illumination. In the early research process, researchers have proposed relatively simple light illumination models such as the Lambert diffuse reflection model and Phone model [17], as well as more complete light illumination models such as the Blinn model and the Cook-Torrance model. The overall light illumination model has been proposed, such as the Whitted model and Hall model, as well as line and area light sources. After the 1980s, the ray-tracing algorithms and radiance method were introduced to create a new field for the research and development of light illumination models [18].

In terms of light source design, based on the study of the physical characteristics of light, this article simulates a light illumination model suitable for ocean scenes based on relevant optical theories. A spherical (simulated sun) luminous body is added, and the light source is oriented in all directions in space. The brightness of the light is evenly distributed, limited to the ocean scene during the daytime. In the experiment, a large cube was created as the main scene of this experiment (Fig. 6). We used atomization, particle system, other methods to simulate the space of the seafloor. In the interior, the terrain system was used to simulate the sea bottom terrain to build a complete scene and try to restore the seabed light as real as possible.

In terms of considering fish habitat or reproduction, we use the form of big fish eating small fish to control the number between fish schools. When encountering various

Fig. 5. Small shipwreck reef

Fig. 6. Undersea simulation

situations, we have done some simulation action research, such as birth rate, mortality, escape rate, to achieve the purpose of simulating ecological changes. Besides, in order to explore the influence of the maximum visual ability range on the fish school structure, random coefficients are added to the fish school mathematical model to maintain a certain degree of chaos to avoid abnormal and orderly running processes (Fig. 7). Consider the impact of external conditions: This simulation study is to interact with individuals in a group without interference from outside predators.

Fig. 7. Random movement of fish

5   Conclusion

The marine tourism consumption market is vast [19], with enormous economic and social benefits. With the popularization of virtual reality tourism, tourists experience a new type of tourism. The integration of virtual reality into the marine ranch is conducive to broadening the marine tourism business model, breaking the traditional preaching and text-browsing cultural spread, breaking through the limitations of today's marine ranch research scope. The combination of the two can not only improve the marine ecological environment and tourism quality but also have the functions of marine cultural penetration and risk aversion.

Therefore, a brief discussion on the future development trend based on this experiment can be expanded into the following three application fields:

(1) In the field of marine environmental protection and ecological restoration, virtual marine ranch is used to promote ecological recovery, build a national research platform for marine ranch, and carry out relevant institutional mechanisms, such as cooperative mechanism for marine ecological environment protection; research on environmental pollution prevention and remediation technology and mechanism in the process of island reef utilization. To provide technical and institutional guarantee for the sustainable utilization of marine resources.

(2) In education tourism, accelerate the transformation of scientific research results. The focus is to use the development prospects of virtual marine ranch technology, innovate education methods, enhance user learning interests. Construct an interactive, sharing, open, and innovative three-dimensional tourism-based teaching model to provide strong support for the marine resource ecological information display and promotion and application.

(3) In terms of intelligent construction, with the help of virtual reality, cloud computing, big data, and other technologies to promote the construction of a smart ocean, the virtual marine ranch can not only bring convenience to teaching and research work,

but also help to ensure the safety of R&D personnel, break the constraints of time and space, and improve work efficiency.

References

1. Yan, H.: Study on the ecological environment evolution and evaluation of the marine ranch tourism area of Wuzhizhou Island. Sanya. Hainan University (2017)
2. Moberg, O., Salvanes, A.G.V.: Ocean ranching. In: Encyclopedia of Ocean Sciences. Academic Press, pp. 449–459 (2019)
3. Choi, Y.B., Boo, C.S., Kim, M.C.: Tourism resources and development plan of marine ranch of Chagwido. J. Fish. Mar. Sci. Educ. **24**(3), 378–386 (2012)
4. Liu, Y., Guo, C., Sun, C., et al.: Construction and practice of virtual ship interactive experimental platform in the area of internet. Lab. Res. Explor. **37, 175**(10), 162–167 (2018)
5. Morton, B., Blackmore, G.: South China Sea. Mar. Pollut. Bull. **42**(12), 1236–1263 (2001)
6. Burdea, G.C., Coiffet, P.: Virtual Reality Technology. Wiley, Hoboken (2003)
7. Cattano, C., Fine, M., Quattrocchi, F., Holzman, R., Milazzo, M.: Behavioural responses of fish groups exposed to a predatory threat under elevated CO_2. Mar. Environ. Res. **147**, 179–184 (2019)
8. Noleto-Filho, E.M., Pennino, M.G., Gauy, A.C.D.S., Bolognesi, M.C., Gonçalves-de-Freitas, E.: The bias of combining variables on fish's aggressive behavior studies. Behav. Process. **164**, 65–77 (2019)
9. Zhu, X., Ni, Z., Ni, L., Jin, F., Cheng, M., Li, J.: Improved discrete artificial fish swarm algorithm combined with margin distance minimization for ensemble pruning. Comput. Ind. Eng. **128**, 32–46 (2019)
10. Hubbard, S., Babak, P., Sigurdsson, S.T., et al.: A model of the formation of fish schools and migrations of fish. Ecol. Model. **174**(4), 359–374 (2004)
11. Breder Jr., C.M.: Equations descriptive of fish schools and other animal aggregations. Ecology **35**(3), 361–370 (1954)
12. Parrish, J.K., Viscido, S.V., Grunbaum, D.: Self-organized fish schools: an examination of emergent properties. Biol. Bull. **202**(3), 296–305 (2002)
13. Kruk, A., Lek, S., Park, Y.S., et al.: Fish assemblages in the large lowland Narew River system (Poland): application of the self-organizing map algorithm. Ecol. Model. **203**(1–2), 45–61 (2007)
14. Zhang, C., Zhang, F., Li, F., et al.: Improved artificial fish swarm algorithm. In: 9th IEEE Conference on Industrial Electronics and Applications. IEEE, pp. 748–753 (2014)
15. Seaman, W., Lindberg, W.J.: Artificial reefs. In: Encyclopedia of Ocean Sciences. Academic Press, pp. 226–233 (2009)
16. Wang, Q., Yan, H., Xu, F.: Overview of artificial reef construction. Agric. Technol. Serv. **34**(03), 149–151 (2017)
17. Stavn, R.H., Weidemann, A.D.: Optical modeling of clear ocean light fields: Raman scattering effects. Appl. Opt. **27**(19), 4002–4011 (1988)
18. Du, Y., Zhang, X., Li, W.: A lighting model for marine scenes in VR system. J. Harbin Eng. Univ. **3**(04), 19–21 (2001)
19. Orams, M.: Marine Tourism: Development Impacts and Management. Routledge, Abingdon (2002)

A Virtual Reality Training System for Flood Security

Rui Dai[1], Zizhen Fan[1], and Zhigeng Pan[2,3(✉)]

[1] Zhejiang Tongji Vocational College of Science and Technology, Hangzhou, China
[2] NINED Digital Technology Co., Ltd., Guangzhou 510000, China
zgpan@hznu.edu.cn
[3] Guangdong Academy of Research on VR Industry, Foshan University, Foshan, China

Abstract. As we can see in the real world, there are frequent public security accidents (earthquake, fire, virus, and flood) in the past years. When the ordinary people are facing the flood disaster, they often do not know how to avoid danger and save themselves. By using the most advanced VR technology, we can popularize public safety knowledge, and improve public security awareness. This paper first describes the related works in flood simulation modeling, VR training, and interacting techniques in VR systems. Then it discusses the implementation of a virtual training system for flood security education based on Unity3D engine which is widely used for the time being. The users can not only experience the horror of public safety disasters in the virtual world, but also learn how to save himself in the event of flood disasters in the virtual environment.

Keywords: Public security · Flood simulation · VR · HCI

1 Introduction

As we all know, there are frequent public security accidents (fire, virus, and flood) in the past years in the real world [1]. Especially in China, there are many floods in recent years or every year. One serious example is that happened in 1988. Figure 1 are some scenes home and abroad.

Generally speaking, the public security awareness of Chinese nation is poor, and they often suffer heavy losses in the face of flood, earthquakes, fires and other disasters. We will only discuss the issue of flood in this paper. It has become the top priority of public safety work to carry out basic public safety knowledge popularization of the general public, actively carry out corresponding exercises and practical drills, and improve people's ability to cope with public security incidents.

The public security consciousness of our people is generally scarce. To solve this problem, we can use virtual reality to do the virtual training, so that ordinary people can see the simulated results of flood, and know the method for escaping from the flood. To implement the virtual training system for flood security, one important component or key technique is the modeling of flood simulation, which will be described in the second section.

© Springer-Verlag GmbH Germany, part of Springer Nature 2020
Z. Pan et al. (Eds.): Transactions on Edutainment XVI, LNCS 11782, pp. 126–134, 2020.
https://doi.org/10.1007/978-3-662-61510-2_12

(a)   (b)

(c)

Fig. 1. Typical examples of flood home and abroad

1.1 Flood Simulation

Floods have caused widespread damage throughout the world. Modelling and simulation provide solutions and tools which enable us to forecast and make necessary steps toward prevention [2]. Thus it is very important to design and implement flood simulation systems. Actually there are many systems have been developed in the past 20–30 years.

Recent research work has shown that flood risk continues to escalate globally, despite an increase in the primary outcome of flood simulation systems. This means that we need employ the related knowledge in the specific field while the flood simulation systems are implemented [14].

There are early demos generated by related system that the goal can be accomplished when an urban flood model is used as a tool. Thus it can get together local lay and scientific expertise around local priorities and perceptions [14]. We will introduce those tools and related techniques in the following section.

1.2 Virtual Reality Training System

Based on the investigation we made in previous research work, we find that, there are several main method for popularizing public security knowledge [3]:

(1) School education. This public safety knowledge can be arranged in the courses in primary or high schools.
(2) Lectures: A series of public safety knowledge lectures can be held in public library or in universities.
(3) Public media: Popularization through media tools such as public safety knowledge manuals, videos and TV programs.
(4) Professional training centers: In some modern cities, there are professional centers, which have some installations which illustrate the public safety knowledge.

These methods have been widely used in many countries, and obtain good effects. However, there are many limitations, which can be resolved by using the virtual reality (VR) techniques. In the three-dimensional virtual environment constructed by VR technology, users interact with objects in the virtual environment in a natural way, greatly expanding the ability of users to understand the real world. The VR system is an extremely complex system involving graphics and image processing, speech processing, audio recognition, artificial intelligence, user interfaces, sensors, multimedia databases, system modeling, etc. [4]. There are many tools for developing VR training systems, a typical tool example is Unity and a typical setup of VR training system is shown in Fig. 2.

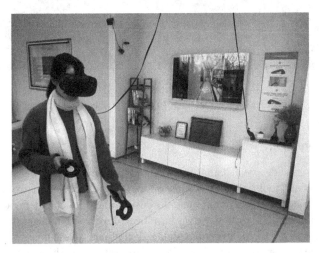

Fig. 2. A typical setup of a VR training system

1.3 Human Computer Interaction Virtual Reality Training System

In VR system, human-computer interaction (HCI) has been an important research focus, the way of HCI has also been developed very quickly. The final purpose of HCI is to achieve natural interaction. In other words, only by the acquisition of knowledge and experience in daily life and without learning or just little learning, the users can manipulate and control the objects in the virtual environment [5–7].

According to the classification of the input device, gesture recognition can fall into three categories: instrumented glove based method, color marker based method and vision based method [8].

(1) Instrumented glove based method: the instrumented data gloves with sensors are worn to obtain the position and direction of fingertip and palm [9]. This method can capture timely and accurately the position and orientation of the palm and fingertips [10].
(2) Color marker based method: users are required to wear color gloves and each finger in the color glove has different color from others [11], which makes it easy to segment out every finger clearly with the help of color segmentation method [12].

(3) Vision based method: bare hand is employed to extract the hand region required for hand gesture recognition. The advantage of this approach is that it is very natural and it directly interacts human with the computer. Gesture recognition based on the FEMD-TMM shows promising performance in vision based method [13].

2 Flood Simulation Modeling

The occurrence of flood events has become more frequent in many parts of the world over the past 30 years and, consequently, more people are exposed to flood damages. Modeling and computational simulation provide powerful tools which enable us to forecast, in order to reduce flood damages. This is very important in China and other countries. Thus a lot of research and development work are carried in home and abroad.

As we all know, the numerical simulations are an important component of flood studies. The amount of input data used for numerical simulations can be very large, both in the number of files and the amount of data required. This is particularly true if numerous simulations are created in order to study multiple scenarios. Zundel [14] described an approach for sharing model data between simulations and a graphical user interface which has been created to manage the simulations and data sharing. The method is based on the previous research and development work. Actually these methods are employed in the product development of the TUFLOW company [15], while these products are widely applied in real situation of different countries.

Based on the review of related research and development work, we found that TUFLOW [15, 16] is a finite difference based 1D/2D hydraulic model.. The result of A TUFLOW simulation consists of all of the data for a single model run which includes 2D grid, 1D segments, boundary conditions and model parameters and options [16], which can be seen in the Fig. 3.

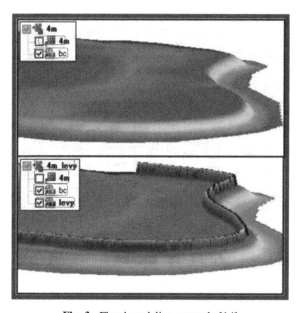

Fig. 3. Flood modeling example [14]

Based on the introduction in reference [14], the picture shown in Fig. 4 is divided into two parts which represent the elevation output from TUFLOW. The grid components are shown in the top left corner of each part of the image. The top grid component (4 m) includes a link to the grid named "4 m" and a link to the GIS (Geographic Information System) layer "bc," which provides boundary conditions. The grid component named "4m_levy" has links to the same grid and to the GIS layer "bc" but also includes a link to a layer named "levy." The levy layer includes the geometry definition of the levy which overrides the elevations in the base geometry.

Also there are some other extended works based on existing system, and the functions are more complete. As shown in Fig. 4, the system is developed by Lee [17]. The system can simulate different results with different mechanisms and parameters, while different colors are employed to indicate detailed information.

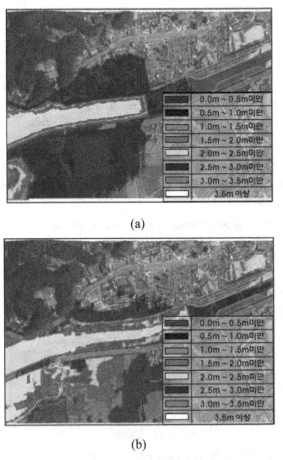

(a)

(b)

Fig. 4. Effect of different flood resolution mechanisms [17]

Of cause, there are some other methods in this field. For example, A typical research work in this field is by Gaudiani [18]. This work addresses input parameter uncertainty toward providing a methodology to tune a flood simulation system (also called simulator) and achieve lower error between simulated and observed results. Based on their experiments, the method is tested at the lower basin of the Paraná River in Argentina (see Fig. 5 [18]), and the percentage of improvement over the original simulator values ranges from 33 to 60%.

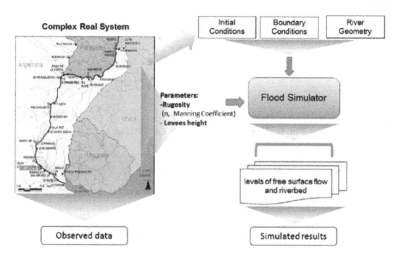

Fig. 5. Case study for the Paraná River in Argentina [18]

3 System Design and Implementation

Based on the previous discussion and the reviewing of the related work, we design the planned system, and present the implementation diagram. We employ uses VR and human-computer interaction techniques to design and implement a virtual training system for public safety education, especial used for flood. The scenes of the system can be produced by 3D MAX software or other modeling software. And the scene models are imported into the Unity3D engine for scene construction and rendering. Finally, we uses HTC VIVE for displaying the 3D scenes in head set. For the navigation and interaction, two modes are designed, the handle provided by HTC VIVE and gesture based on bare hand.

We discussed the interaction of the first mode in the following, and in the future we will implement two modes. In such a VR training system, the users can operate the handle to move and interact in the virtual scene, as if he is viewing and walking in the real world. We divide the system into several modules shown in Fig. 6, which shows the overall architecture of the designed system.

Based on the design diagram, we will implement the panned system in the coming months. In order to let the users understand what the flood, the cause of the flood and how to save themselves from the flood, the system will contains the following parts:

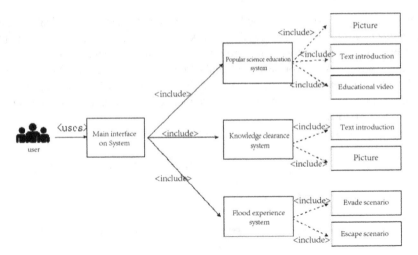

Fig. 6. Diagram of the system architecture

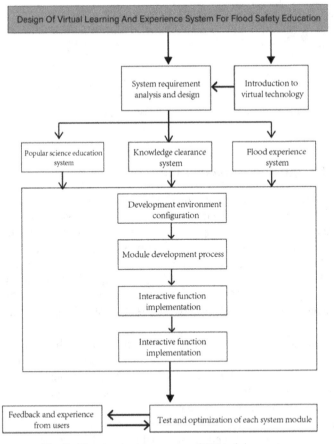

Fig. 7. The developing process of VR training system

(1) A science education room. In this virtual room, the users will learn the basic public safety knowledge, especially in the theme of flood.
(2) An immersive room for obtaining the experience of flood.
(3) The game room for planing games related to escaping from flood.

The actual implementation diagram of the designed is shown in Fig. 7. In the designed system, it will support the simulation of flood in two dimensional mode and in three dimensional mode. Examples are shown in Fig. 8, which are implemented in other systems. We will aim at the goal and obtain the similar results.

(a)   (b)

Fig. 8. The simulation of flood of other systems, which will be implemented

4   Conclusion and Future Work

This paper designs and implements a VR training system for public safety education with the theme of flood. The related work (flood simulation modeling, virtual reality training system, and human-computer interaction) are described first, then the design diagram and implementation schemes are discussed, which will be used as the guideline for implementation in the near future.

As we point out in the first section, that at present citizens' awareness of safety is low, safety knowledge is generally scarce, and the method of popularizing safety education is not effective. For the future work, first we need implement the system step by step, then we will test the system, asking some users to evaluate the system, and improve the system based on their comments and suggestions.

Acknowledgments. This research is supported by the project of the Bureau of Water Science & Technology in Zheijiang Province (grant number: RC1970). Also, we acknowledge the support of the Guangzhou Innovation and Entrepreneurship Leading Team Project under grant CXLJTD-201609.

References

1. Huang, X., Bai, H.: Risk prediction of rural public security environmental carrying capacity based on the risk entropy. Nat. Hazards **90**(1), 157–171 (2018)
2. Pan, Z., Chen, J.: VR-based edutainment. Virtual Real. **12**(1), 1 (2008)
3. Pan, Z., Zong, Y.: Virtual training and experience system for public security education. In: Pan, Z., Cheok, A.D., Müller, W., Zhang, M., El Rhalibi, A., Kifayat, K. (eds.) Transactions on Edutainment XV. LNCS, vol. 11345, pp. 229–237. Springer, Heidelberg (2019). https://doi.org/10.1007/978-3-662-59351-6_15
4. Fischbach, M., Wiebusch, D., et al.: Semantic entity-component state management techniques to enhance software quality for multimodal VR-systems. IEEE Trans. Vis. Comput. Graph. (2017). https://doi.org/10.1109/TVCG.2017.265709
5. Wachs, J.P., Kelsch, M., Stern, H.: Vision-based hand-gesture applications. Commun. ACM **54**(2), 60–71 (2011)
6. Ibraheem, N.A., Khan, R.: Survey on various gesture recognition technologies & techniques. Int. J. Comput. Appl. **50**(7), 38–44 (2012)
7. Lee, B., Chun, J.: Manipulation of virtual objects in marker-less AR system by fingertip tracking & hand gesture recognition. In: Proceedings of the 2nd International Conference on Interaction Science, pp. 1110–1115. ACM, Seoul (2009)
8. Guo, S., Zhang, M., Pan, Z., Sun, M.: Gesture recognition based on pixel classification and contour extraction. In: Proceedings of International Conference on Virtual Reality & Visualization, 01 October 2015
9. Garg, P., Aggarwal, N., Sofat, S.: Vision based hand gesture recognition. World Acad. Sci. Eng. Technol. **49**, 972–977 (2009)
10. Pratibha, P., Vinay, J.: Hand gesture recognition for sign language recognition: a review. Int. J. Sci. Eng. Technol. Res. (IJSETR) **4**(3), 466–467 (2015)
11. Lamberti, L., Camastra, F.: Real-time hand gesture recognition using a color glove. In: Maino, G., Foresti, G.L. (eds.) ICIAP 2011. LNCS, vol. 6978, pp. 365–373. Springer, Heidelberg (2011). https://doi.org/10.1007/978-3-642-24085-0_38
12. Dhruva, N., et al.: Novel segmentation algorithm for hand gesture recognition. IEEE, pp. 383–388 (2013)
13. Ren, Z., Yuan, J., Meng, J., Zhang, Z.: Robust part-based hand gesture recognition using kinect sensor. IEEE Trans. Multimed. **15**(5), 1110–1120 (2013)
14. Zundel, A.K.: Flood modeling simulation management. Technical report, Brigham Young University (2006)
15. Syme, W.J.: Dynamically linked two dimensional/one dimensional hydrodynamic modelling program for rivers, estuaries and coastal waters. M.Eng.Sc. thesis, University of Queensland (1991)
16. WBM Pty Ltd. TUFLOW users manual. WBM Oceanics, Australia (2005). http://www.tuflow.com/Downloads_TUFLOWManual.html
17. Lee, H.-J., Hong, S.-H.: Flood simulation by using high quality geo-spatial information. J. Korean Soc. Geospatial Inf. Syst. **18**(3), 97–104 (2010)
18. Gaudiani, A., Luque, E., García, P., et al.: Computing a powerful tool for improving the parameters simulation quality in flood prediction. Procedia Comput. Sci. **29**, 299–309 (2014)

Life Science Immersive System Based on Virtual Reality

Yu Gao, Yin Wang, Yihao Chen, Jie Chang, and Shouping Wang[✉]

School of Data Science and Software Engineering, Qingdao University, Shandong, China
{clamsama,wywangyin1998,elianacc2019,changjie1997}@outlook.com,
wangshouping@qdu.edu.cn

Abstract. This paper presented a virtual reality experience system based on the theme of human body and cold virus struggle. In order to construct a realistic human cell scene, we applied physic-based rendering game production and Unreal® Engine. The proposed project used virtual reality, combining educational theory. Plentiful functionalities of virtual system were discussed in detail. As such, users could well participate in a war of cells to against virus from different perspectives. It is thought that users can become active learners, with they might well taking the initiative to acquire scientific knowledge and summing up their thinking in a lifelike experience, ultimately cultivating the ability of exploration and innovation.

1 Introduction

In various fields of education, the introduction of virtual reality (VR) has grown in recent years. VR technology can provide users a simulated experience with a first-person perspective via various sensory stimulations [1].

Life science is closely related to everyone's life, for instance, biological functions, regulatory mechanisms and evolution laws of the living system, happens anytime in our bodies [2].

VR system could well build a biological environment, which is extremely difficult to construct in reality, in turn it potentially being regarded as a promising teaching science tool. In the University of New England in Australia, an educational project using immersive VR software aims to teach empathy regarding the aging process to health profession students by leading the learners through a simulated aging experience (e.g. macular degeneration and hearing loss) [3].

This project systematically integrates virtual reality and the theory of life science theory, illustrating a bioscience-themed scene. By using Computer stereoscopic modeling technology and UE4 (Unreal® Engine 4), we constructed a scene of cells living in our bodies, and this could allow users explore realistic three-dimensional scenes and take part in a real war with the virus in a virtualized way.

Compared with other software which only provides piecemeal knowledge, this project offers users an integral virtual setting. With the help of human-computer interaction, users can enjoy a transforming experience. As such, they arguably turn abstract scientific theory into practical experience of the imaginary, which means they ultimately

© Springer-Verlag GmbH Germany, part of Springer Nature 2020
Z. Pan et al. (Eds.): Transactions on Edutainment XVI, LNCS 11782, pp. 135–145, 2020.
https://doi.org/10.1007/978-3-662-61510-2_13

can actively acquire the scientific knowledge, in turn summing up the thinking in the life-like experience and broadening their knowledge through the systematic science and software.

2 System Design

2.1 Plot Design

Project selects a topic of cold virus intrusion, which includes three parts: structure of cold virus, process of cold virus invading the human body and struggle between human cells and cold virus. These processes are shown in different models, which not only systematically introduce the relevant information, but also vividly display the untouchable knowledge to help users' learning initiative and innovation (Fig. 1).

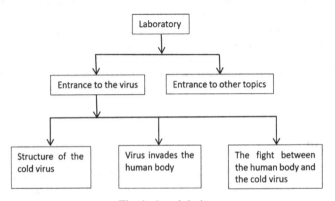

Fig. 1. Level design

2.2 Level Design

Main interface is a laboratory scene, providing a representative visualized object for each topic placed in different locations of the lab. Research direction will be placed on a test-bed in the form of holographic projection (Fig. 2).

Fig. 2. Laboratory scene   **Fig. 3.** Virus structure

When user clicks on the virus model, it will enter the secondary interface - human cell environment, as shown in Fig. 3, in turn three physical buttons being placed on the secondary interface: cut-virus section model, virus itself and leukocyte, which allow users to click on the corresponding physical objects to enter different scenes. As such, they could well watch the structure of the virus and the process of virus invading human body, and taking action and participating as virtual guards to help white blood cells defeat virus (Figs. 4 and 5).

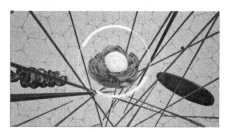

Fig. 4. Virus invading cells   **Fig. 5.** Leukocyte phagocytic virus

2.3 Interactive Design

The software interaction is based on Oculus Rift head-mounted display. Basic interaction includes role movement, scene jumping, scene model interaction with characters, animation navigation, person switching, etc., and above interactions can be completed through the handle of virtual reality device.

2.4 Animation Design

The software animates include three parts, including: human environment, animation navigation and particle effects. The former includes simulated movement of viruses and cells, virus invading cells and human cells phagocytizing viruses. Scene model part is made in 3D Studio Max and the animation is done in UE4. Traditional CG (Computer Graphics) particle effects are produced by software such as n-Particle and Mental Ray, during its render process, projects firstly solve the solution, then output the rendering effect. Although the quality of picture is better, it cannot perform real-time rendering operations or impose a high load on the computer system. Unreal 4 engine provides a tool to create sophisticated models and effects. Its powerful cascading particle system could well create eye-catching renderings and lifelike particle effects [4].

3 Design and Implementation

The software design implementation includes three parts, model building, interactive UI and animation.

3.1   Construction and Import of 3D Models

3D models needed for the project include persona, laboratory, virus, leukocyte, red blood cell, organelles, intracellular environment, human cell tissue environment, etc. The construction of 3D scene and object includes the production of 3D model and expression of textures. When constructing 3D scenes and objects, we should show high quality picture effect and take into account the running frame rate of VR screen. The action area of user is planned in an early stage, which means that the high-precision model should be used in close-range object, furthermore, the distant object can use low-precision model, even a map to represent. The following is a brief description of the construction steps of core biological models in 3D Studio Max software, and some matters needing attention while we import modules.

Virus Model: The Virus produced in the project is adenovirus, which is the culprit of acute respiratory tract infection. Acute respiratory infection is one of the most common disease, which spreads widely all over the world and potentially induce influenza-like symptom diseases at the same time. Influenza-like symptom diseases are the second leading cause of death in children. Up to 90% of children with pneumonia are infected by the virus. Adenovirus can replicate and proliferate in human cells, then the host cell cleaves and dies, invariably release a large number of progeny viruses, in turn resulting in further infection, which can do is induce a variety of cancers [5].

Firstly, we should create a new polygon in 3D Studio Max software, converting it to editable mode, then selecting all the edges in the edge interface. Subsequently, checking the NURMS subdivision and converting it to a polygon again, next deleting all the edges at this time, with Some hexagons and pentagons appearing in this step. Thereafter selecting several hexagons and pentagons and using NURMS segmentation. Finally, using the material editor to add a material map.

Protein Model: A geometric sphere in the basic body of 3D Studio Max should be created in first step, then opening the particle system and creating a particle array. Next Clicking pick object to pick up the geometry sphere, in order to make a particle emitter. Then selecting the gridding command to create gridding and picking up particle objects. At the same time, changing number of particles and the speed of particle motion to adjust the shape of the particle. Finally, scaling and adjusting it to complete the basic framework of the protein.

Red Blood Cell Model: The first stage involves creating a cylinder and set its radius, height, segments, and check the smoothing option. Later checking the upper and lower faces of the polygon, select insert, chamfer, collapse, and use NURMS subdivision.

Secondly, selecting the noise in the modifier list and setting the scale, which not only makes the red blood cells look less smooth, and improves the texture of the model. Finally, adding color to the red blood cells.

Considerations for Importing the Environment
Import Model
Most of models exported by 3D Studio MAX and Maya software invariably are in FBX format. When importing the FBX file into Unreal 4, we should consider that the property

of model should be carefully adjusted and further modified, such as adding rigid bodies and modifying collisions. As such, the skeleton mesh should be checked in the import option.

Model Blackened After Rendering

When Unreal 4 renders a model with its original material, the problem of blackening the model typically occurs. This problem arguably caused by the overlap of the illumination UV (texture coordinates) of the model. So be sure to expand the model UV without overlap in 3D Studio MAX or Maya. Slender model may require additional lightmap resolution in Unreal 4.

Physical Volume

Actor, the entity in scene, has a Blocking-Volume button in their detail panel, which allow us modify physical volume and other properties. Physical volume, however, is not visible in the game, but real exist. It can be used with various components, to complete event triggering and other operations. By clicking on the Blocking-Volume button, we can create a separate collision box for the current Actor. Furthermore, The Alt + C key combination can hide or show the collision box.

3.2 Mode and Character Control

Basic Elements of the Project: Game Mode, Player, HUD Controller

Once the model and scenes were created and imported into the Unreal 4 engine, the production project could be started. Firstly, we should build a basic circumstance in the engine.

One of the biggest advantages of the Unreal Engine is the native support for C++, one of the most widely used programming language worldwide, from trivial commercial automation software to heavy data processing in the search for new algorithms in the field of computer science [4]. Due to its traditional and sophisticated method, C++ was also widely applied in large-scale game development and data processing. However, Unreal Engine provides another visual programming language: Blueprint (BP), which is versatile visual script that allow users to make virtual program development and statement writing easier. The following are specific step instructions to construct a fundamental operate circumstance.

Firstly, opening the level in the Unreal 4 engine and creating some empty folders, in order to hold the BP files and role-related files. Then creating a new Blueprint, as the game mode and typing selection Game Mode, named VR Mode.

In the second stage, we should create a new HUD blueprint. Head Up Display is the full name of HUD, HUD should be used as the controller. As such, it will be selected as Player Controller in VR.

Ultimately, creating a new blueprint as our First Player Character (1P), then we need specify the default character and mode of the game. In the right property bar, we can bind the HUD which just created. After this, specifying the default game mode of the engine should be defaulted in the project settings. What need we do is select Project

Settings, in the upper left Edit menu, then specify the game mode of the project under Maps & Modes, as the VR Mode we created.

Input Response and Import Role
In this step, we will import the role model into UE4 project and implement the key response of the role. There will be a slight difference between importing a character and a model. Since the skeleton will be accompanied by the imported character, we should not check Textures option and Materials option. After importing, there will have three resources, character model, physical resource and the skeleton. In order to ensure material used by the bone file is only deployed for the bone file, we need to create a new Material ball as the character's material, in turn corresponding options are automatically opened at the bottom of the material.

Subsequently, opening Project settings under editing, at the same time, inputting menu of the engine and controlling the operation of the mobile class in order to generally invoke Axis Mappings, such as moving forward or turning the view. Action Mappings generally controls the common release spells or jump dodging in games.

We can firstly make keyboard and mouse input response, which is easy to test in UE4 preview mode. The input response method of virtual reality device is the same as that of keyboard and mouse, in turn bounding the handle button, rocker, and events (Fig. 6).

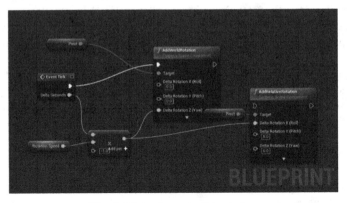

Fig. 6. Blueprint operation window

3.3 UI

Virtual reality could well give users a stronger visual experience. It is also extremely easy to make people feel out of step to see the traditional UI in VR environment. The project materializes the traditional information such as menus, buttons, and text, which not only makes the interaction interesting, but also makes the user more immersive.

Custom Cut Out Menu and Exit Menu
Menus are indispensable in a variety of applications. Through the project menu, users

can perform interactive operations such as custom settings, pauses and exits. We can use the Unreal 4 engine control blueprint to create a control menu with basic functions.

Manual Open Menu
In the process of testing program, we will see that the runtime menu is enabled by default. We have to change the menu to open by pressing the button. Firstly, changing the Visible property to Hidden so that the UI is hidden by default. Then clicking Open Is Variable and change the layer to a variable so that we can dynamically modify it in blueprint.

Restrict Operations When Opening Menus
We need ensure that the player can't move at will after opening the menu. To achieve this function, we can use Boolean variables to record whether the player has opened the menu, in turn changing the variable property to true when player triggers it. Then checking the menu to exit the game. Clicking the On Clicked, which under the Events menu, to create a click event. Furthermore, creating a click event for each button. Finally, creating a new Print String node and calling the output character to view function to detect whether the click event is valid.

Implementation of Perspective Switching in Menu
When using virtual reality software, the immersive first-person perspective is used, the third-person over-the-shoulder perspective (3P) can simultaneously view a wide range of environments, and the lens shift is moderated, which is suitable for 3D dizzy people. Below we will set the function of the person to switch in the menu.

Create Interface
Creating a new interface as a bridge to the blueprint. Firstly, creating a new function, named SwitchPawn1P as the name of the function to switch to 1P. Meanwhile, creating a new function, named SwitchPawn3P as the name of the function to switch to 3P. Then copying the function name for easy searching and use the created blueprint interface. Next, creating a new node Get All Actors with Interface and calling to search all objects by using the blueprint interface in the drop-down list. Selecting the created blueprint interface so that we can automatically search for objects using the created blueprint interface in the scene. Subsequently, creating a new node, Switch Pawn 1P, in order to invoke the ability to send a message. Finally, copying a node that searches for all objects that use the Blueprint interface and produces a message that sends a switch to 3P. Ultimately, creating a new node, SwitchPawn3P, to invoke the function of sending a message.

Logic of Switching:
In order to facilitate the data call, 1P object should be used as a variable firstly, then creating a new variable as a 3P variable and modify it as a 3P object, which can get the player's Pawn functionality and set the runtime Pawn to the 3P variable as 3P. Subsequently, making switch logic and adding a blueprint interface. By using the 1P variable, we can change the current player's type to 1P to test the effect, in turn making the ability to switch from 1P to 3P. It is important to note that when the player moves the position, the position of the 1P birth should also be modified. As such, the location of the 1P birth specified earlier only at the beginning of the game, so the player's location is not bound after the move position. Next, creating a new node that invokes the vector

add function and modifies the value, in order that the 1P birth position is slightly higher than the 3P birth position, then avoiding overlap. Finally, performing copy the node to modify the coordinates and creating a new node Set Actor Transform to modify the connection, then compiling in sequence.

Prevent Role from Resetting:
In preparation for making a feature that prevents the character from resetting, we should add a new variable as a basis for judgment, modify the type to Boolean, and check Is VHPawn 3P. The new node Branch, invokes the function of the branch. Then using Boolean variables to judge whether the current player's variable is 3P, performing the switching function only if the value is true. We next should copy the node, and make a judgment on whether the current player is 1P or not. Finally, using variables to record the state, in turn copying the node and recording the state compilation. After that, returning the scene to check effect.

Interactive Prompts on the Top of the Head

In the process of sightseeing, visual interactive prompts can give users the necessary guidance to enhance the interaction.

During the interactive prompt making process, we can set the visibility function by calling the Set Alignment In Viewport and Set Position In Viewpoint nodes to set the viewport alignment function, so each frame is predicted to execute these nodes after the connection, 2D UI is adsorbed into 3D space, and the function of setting the viewport position is invoked to specify the display position of the UI. The Convert World Location to Screen Location node is also used to invoke the function of world space to plane space. If the compile run does not see the effect, we need to select the collision range model and click entry event, then click to create an event that left, in turn calling the function to remove the drawing. This means that we need to get the variable and tell the node who wants to be removed (Fig. 7).

Fig. 7. Interactive hint

3.4 Animation

Matinee, one of the most powerful UE4 animation edit tool, can be used to create animate and record videos. Its main idea is to provide a set of its own time coordinate system in the Matinee editor, in its relative time axis, by adjusting the properties of actors, we can change the state, in turn achieving the purpose of animation, and this means is that we could well adjust actors, change the state of the camera, the realization of particle effects, the rendering of light, and so on.

Navigation Animation

VR technology now enables demonstrators to create first-person videos that lead observers to experience what nursing educators see and how they move their hands [6]. Navigation animation can help those who first use virtual reality equipment. Users can visit the whole scene under the guidance of the system and freely observe and interact.

By using Matinee and Blueprint, we can make a simple navigation system. Firstly, dragging and dropping a default circle in the scene, as an auxiliary model, in turn navigation being understood as a roller coaster, with user tour that needs to be hidden at run time, thereafter select the model right-click to enter Pilot mode, so that our perspective switches to the currently selected circle.

Secondly, we should adjust the coordinates and angle to the starting point, then click the upper left corner to release the control, with selecting the model and clicking create Matinee to animate and edit. Subsequently, Right-click to add a new empty group and rename it to the CG, then selection group, right-click to select New Movement Track to drag the bottom to extend the timeline, and press Enter to add a key frame at the current point in time. Right-click on the model, with selecting Pilot mode again locking the model to the viewport to move, now that the trajectory is made, ultimately, move to the right position and select Keyframe 2, and so on.

When we make and play animations in Blueprint, the user can freely observe the effect, similar to the roller coaster, although the route is fixed, users are free to observe the surrounding scenery. Then copy a secondary circle and scale it to 75% to fit above the original model, next drag it under the big circle in the world outline view, in order to set a submodule. We could seem the big auxiliary circle model as a roller coaster, and the smaller ones are tourists. It takes a fixed track. As such, we can sit on the roller coaster.

Ultimately, we need set an event, trigger it when the player was born in the game. Creating a Node Play to play the animation and can be placed in different positions to trigger the tour mode under different circumstances. Subsequently, creating a Node Get Player Controller and Set View Target with Blend, which used to obtain the player controller and transition from 3P to the model. finally, connecting the Play node to the Set View Target with Blend node (Fig. 8).

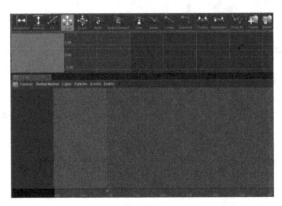

Fig. 8. Matinee window

Free Observation

Navigation animations need to have ability in order to allow users to watch freely and avoid the sense of boring. In this part, we continue to implement in the animation blueprint code.

When the player is playing the animation, initially, we should operate player's input affects. Creating a new Boolean variable to store the control state, meanwhile, right-click on the blank to set the variable and connect it to the Set Actor Hidden in Game, activating when the animation is played.

In the second stage, creating a new node and selecting the previously created coordinate axis event Turn under the input event, then clicking the Boolean variable. In the meantime, right-click to obtain the acquisition variable, which is used as the basis for judgment. Creating node Branch to connect Turn and Boolean variables and copying the name of the visitor model. this is the model we want to control, and we also identify the user who made the roller coaster, then creating a new Add Actor Local Rotation node, calling the function of setting the rotation value of the object itself.

The third stage involves adjusting the maximum elevation angle when free observation, creating a new node Break Rot to call the disconnect function and corresponding to its Min is −60°, Max value It is 60°.

After creating animation, we have to reset the parameters, to allow their normal state. Selecting Matinee and returning to the level blueprint operation interface, then right clicking on the blank space to create a new Matinee controller and creating a new node Set View Target with Blend to set the player's perspective. Finally, we should return to the perspective to the player-controlled character.

3.5 Project Packaging and Testing

One of Unreal 4 engine benefits include "what you see is what you get", which means that we can test our program while implementing project features. Scenarios playback, lighting compilation, blueprint testing and packaging runs are four major parts of the project test. After packaging, an executable program is generated, which is the final version for project development.

During the Unreal 4 package run test, we can change project names and publish the platform and bake. Other options are also allowed in the project Launcher. Project files and directories, which should be emphasized, edit in English as much as possible to avoid some baking errors.

4 Conclusion

By using software such as Unreal 4 and 3D Studio MAX, which have completed all expected functions and achieved expected performances, we could continue to update and improve the functionality of software. Which can do is add interactive chapters of thematic knowledge and immerse users in a fascinating human environment.

Acknowledgements. First of all, I would like to extend my sincere gratitude to my supervisor, Shouping Wang, for her instructive advice and useful suggestions on my thesis. Secondly, I would like to express my heartfelt thanks to the China College Students' Innovative Entrepreneurial Training Plan Program (Grant No. 201811065007), which supported this essay. I am also grateful to my friends, who have been guiding and helping me in the past three years. Ultimately, I would like to thank my dear family for their love and great faith in me for many years.

References

1. Davis, R.L.: Exploring possibilities: virtual reality in nursing research. Res. Theory Nurs. Pract. **23**, 133–147 (2009)
2. Zhou, Y.: Re-exploring the definition of life science and life. Academia Bimestris. Chuansheng Hu. **113**, 112–115 (2010)
3. Dyer, E., Swartzlander, B.J., Gugliucci, M.R.: Using virtual reality in medical education to teach empathy. J. Med. Library Assoc. **106**, 498–500 (2018). https://doi.org/10.5195/JMLA. 2018.518
4. Zhang, B.: Test situation based on unreal 4 engine 3D interactive demonstration system. China Sci. Technol. Panor. Mag. **28**, 27–29 (2018)
5. Wei, M.: Monitoring of viral pathogens in respiratory system in some areas of China. China J. Mod. Med. **13**, 13–44 (2011)
6. Fukuta, J., Morgan, J.: First-person perspective video to enhance simulation. Clin. Teach. **15**, 231–235 (2018)

Augmented Reality Animation for Pancreatic Surgery

Wei Xuan[1], Ling-ling Cui[2]($\boxtimes$), and Xin Lv[2]

[1] Department of Hepatopancreaticobiliary Surgery, China-Japan Union Hospital,
Jilin University, Changchun 130031, China
[2] Jilin Animation Institute, Changchun 130000, China
94640400@qq.com

Abstract. Owing to the complexity of the anatomical structure of the pancreas, surgeons must perform long-term observations and strict training before performing operations. The application of AR technology provides an ideal surgical training platform and has important innovative significance for the surgical training of interns. Through the AR animation of pancreatojejunostomy, training efficiency can be improved. This article focuses on the feasibility and development of AR technology for application to pancreatic surgery, especially the role of AR technology in medical teaching and clinical practice. The study explores the development status, features, design methods, and applications of augmented reality animation as it relates to pancreatic surgery.

Keywords: Pancreatic surgery · Augmented reality technology · Augmented reality animation · Medical application

1 Introduction

Given the complexity of the anatomical structure of the pancreas, surgeons must perform long-term observations and strict training before performing operations to ensure the smooth progress of the surgical procedure, improve the success rate of the operation, and reduce the occurrence of medical accidents [1]. However, mistrust between doctors and patients has intensified medical disputes in China and many patients refuse to allow young doctors to perform surgical operations. Disputes between doctors and patients are becoming increasingly severe. To better guarantee the success of operations, instructors have also reduced the opportunities for interns to perform operations, resulting in their having relatively limited clinical internship experience, which has ultimately led to the relatively unequal or inadequate contact experience of medical students in medical practice.

The application of AR technology would provide an ideal surgical training platform; it provides virtual surgical scenes and clear three-dimensional displays of the surgical site, so a trained surgeon can observe the anatomical structure of the pancreas from any angle and simulate the resection of a lesion, thereby giving a more realistic surgical experience. In addition, owing to the associated repeatability and lack of limitations

© Springer-Verlag GmbH Germany, part of Springer Nature 2020
Z. Pan et al. (Eds.): Transactions on Edutainment XVI, LNCS 11782, pp. 146–153, 2020.
https://doi.org/10.1007/978-3-662-61510-2_14

regarding training time and location, it has important significance for the surgical training of interns and can greatly improve training efficiency.

The design of augmented reality (AR) animation for pancreatic surgery has been primarily based on research into pancreatic surgery. AR animation is a new, interactive, and interesting form of animation. Although it has been applied and developed in some industries, it is still new to most people. The purpose of this article is to enable people to better understand the concept, characteristics, application scope, and design methods of AR animation for pancreatic surgery. The design process involves the creation of more AR animations to enrich pancreatic surgery demonstration cases and the use of the power of new technologies to help more people use AR animations to learn pancreatic surgery techniques, understand surgical principles, and master surgical skills. The significance of this research is to facilitate the application of AR animation to pancreatic surgery, as well as other medical fields, and to present its characteristics.

2   Related Work

2.1   Pancreatic Surgery

Pancreatic malignant tumours are a significant cause of death, given that their onset is hidden, they are difficultly diagnosed in their early stages, they progress rapidly, and present a poor overall prognosis. Because of the limited effect of traditional chemotherapy and radiotherapy, surgical resection remains the main treatment method for pancreatic malignancies. In particular, for the surgical treatment of pancreatic head cancer, whether a traditional open pancreaticoduodenectomy or the currently emerging laparoscopic pancreaticoduodenectomy, the operations present difficulties combined with the high risk of postoperative complications. Further, certain complications are associated with pancreatic jejunal anastomosis. Presently, there are hundreds of pancreatic jejunal anastomosis methods, such as end-to-side duct-to-mucosal pancreatojejunostomy and end-to-end invaginated pancreatojejunostomy, but the accepted anastomosis is still related to the familiarity of the surgeon. The proficiency of the surgeon determines the quality of anastomosis. This study presents a pancreatojejunostomy performed through AR animation, which may make anastomosis methods easier to learn and understand, so as to facilitate the development of clinical work.

2.2   Applying AR to Medicial Engineering

AR technology borrows from virtual reality technology, uses optical projection and real-time computing methods, superimposes information into the user's field of vision, and enhances the user's feeling of interaction with the outside world. It has gradually become a new key technology in the medical field in recent years. With the increasing use of AR technology, more and more broad application prospects are being realised [2]. AR technology can significantly improve the level of medical education, thus reducing the failure rate of operations, improving the accuracy of operations, improving the efficiency of learning, and so on. In medical education, there are higher requirements for teaching methods, students' learning effects, experimental conditions, etc. [3]. The experimental cost of a full-reality scene is high. Although it can be conducted smoothly in some cases, it is easy to lose generality, whereas in an all-virtual environment, it is easy to lose touch with reality [4]. Therefore, AR technology will greatly improve the current situation of medical education, expand the teaching space for teachers into areas that cannot be reached by full reality or previous virtual teaching methods, and provide students with a rich and deep sense of reality and enlightenment so as to greatly improve learning efficiency and reduce experimental and teaching costs. As an example, the pros and cons of the quality of pancreaticojejunostomy are crucial to a patient's postoperative recovery, and the long learning curve in the early stages can cause great inconvenience to patients. Therefore, if the relevant training can be conducted through the use of AR animation designed specifically for pancreatic surgery, the surgeon can quickly master this important step and obtain good results.

3   AR Animation

AR animation uses AR technology for registration tracking and virtual object generation, superimposing animation content onto a real scene, and presenting the animation through devices with sensors and display technology. AR animation is such that the animation not limited to a plane view; rather, it integrates animation with the real world. AR animation is a cross-discipline that well reflects the characteristics of its constituent components. Therefore, the development of AR animation is affected by the development of AR technology.

3.1   Development of AR Animation in Mobile Digital Media

AR animation is becoming increasingly popular in mobile digital media devices, and presents good market prospects. Moreover, AR animation in mobile digital media is also accepted by the public. AR technology, is becoming increasingly mature, with the functions of mobile digital media becoming increasingly abundant. Therefore, the development of AR animation for application to mobile digital media will continue to improve rapidly, such that it will quickly be applied to all fields of society, especially in the field of medicine. With respect to our study, AR animation design for pancreatic surgery will be applied to mobile digital media devices.

3.2   Difference Between AR Animation and Traditional Animation

There are many similarities and differences between AR animation and traditional animation. AR animation is a combination of virtual and real animation that needs to be integrated with AR technology in the production process; this differs from the traditional AR animation, which uses two-way communication. Not only can the system convey information to the audience, the audience can also interact with the characters in the animation. The playback of AR animation is not limited by the screen size, and the audience can view the animations in 360°. AR animation also differs from traditional animation in terms of media, mode, and application.

AR animation is characterised by its combination of virtual and real images, entertainment and interaction. AR animation has been applied to many industries.

3.3 Approach to Spread AR Animation

The development of AR animation based on mobile digital media has proven to be feasible. Mobile digital media devices such as smartphones and tablets with high computing powers and camera and display capabilities are capable of fully displaying AR animation. With a smartphone or tablet, users have a basic AR display system. This is a method of communication closely related to this topic. In addition, large-screen interactive platforms and AR glasses designed to be worn by a user can also display AR animation.

4 AR Animation Design for Pancreatic Surgery

To complete the perfect superposition of an animation and a real scene, three steps are required: the first is tracking and registering the animation and the real scene, the second is the virtual generation, and the third is displaying the AR animation. Presently, augmented animation based on mobile digital media is a tag-based AR, approach, and it is also a common method to view AR animations in our life. In an actual application, an AR animation combining virtual and real images is identified, marked, and displayed on a display device through the use of sensors.

Through the AR animation of pancreatojejunostomy, students can better understand the technology of marker-based AR technology and marker-based AR animation. To achieve identity-based AR animation, we must first acquire an image and generate a virtual object. As the camera of the display device acquires an image, an interactive superposition of the acquired image and the operation image is created, allowing an image of the operation to be displayed. This display process involves two steps.

First, logo images are created. Vuforia provides a series of tools for creating objects, managing object databases, and managing the program licenses. The image to be identified is uploaded to Target Manage for use on AR devices and in the cloud. The Unity Editor option is selected in the Download Database and the obtained file is imported into Unity. The AR animation of pancreaticojejunostomy in pancreatic surgery is described in detail. A logo image of pancreaticojejunostomy is created, as shown in Fig. 1.

The second step in the AR animation process involves virtual objects being generated and displayed; virtual objects can be converted into a three-dimensional model using Autodesk Maya, Autodesk 3ds Max, and other modelling software. The pancreas and jejunum in the case of pancreaticojejunostomy are modelled in Autodesk 3ds Max, as shown in Fig. 2. Next, the texture of the pancreas and jejunum are drawn, as shown in Fig. 3, and the pancreas and jejunum are animated, as shown in Fig. 4. They are then mapped, as shown in Fig. 5. The material of the pancreas and jejunum are mapped, as shown in Fig. 6. The animation of the pancreaticojejunostomy model is thus completed.

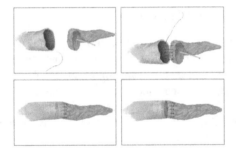

Fig. 1. Acquisition of image and creation of virtual image for pancreaticojejunostomy.

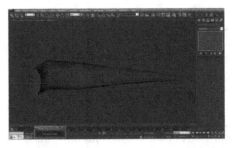

Fig. 2. Models of the pancreas and jejunum made in Autodesk 3ds Max.

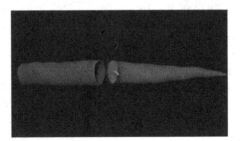

Fig. 3. Texture drawing of the pancreas and jejunum.

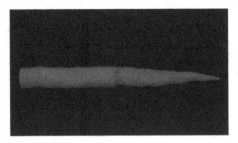

Fig. 4. Animation of the pancreas and jejunum.

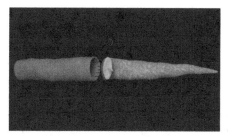

Fig. 5. Mapping the pancreas and jejunum.

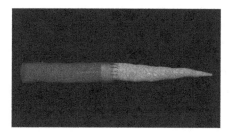

Fig. 6. Creating the material of the pancreas and jejunum.

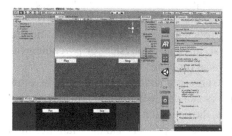

Fig. 7. Importing the model animation in Unity.

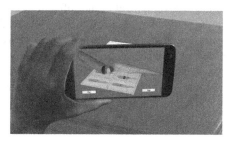

Fig. 8. Scanning and identifying the logo image (Fig. 1) to show the effect of AR animation.

The animation of the pancreaticojejunostomy model is imported into Unity, and the C# programming language is used to make the button function. Finally, an APP file for Android or IOS is exported, and the logo graphics are scanned with the phone camera (Fig. 1) to facilitate the AR animation. As a result, we can view the entire pancreaticojejunostomy process via a 360-degree AR animation, as shown in Figs. 7 and 8.

By successfully performing AR animation design for pancreaticojejunostomy, we can complete significant amounts of AR animation design and production for pancreatic surgery, which can be used in medical teaching, surgical training and other fields. This will help improve the quality of medical education and clinical practice skills.

5   Application of AR Animation in Medicine

Applying AR animation to experimental medical education has unique advantages for the cultivation of the skills of medical professionals: it can simulate difficult teaching scenes, making them visual and participative. Through AR devices, users can experience animated and interactive pages in medical textbooks in an almost completely virtual environment [5].

AR technology combined with animation design can be used as a platform to allow medical students to repeat exercises in a risk-free environment far away from patients, so that medical students and junior doctors can observe and analyse cases from multiple angles [6]. In a virtual environment, they can operate repeatedly for a surgical procedure, enhance their technical proficiency, and greatly improve the probability of successful operation, without the risk of causing physical or psychological harm to patients and encountering medical ethical problems [7, 8].

Through AR virtual surgery operation combined with animation design, medical students and junior doctors can gain perceptual knowledge and personal experience of an operation; additionally, the operation plan and main process designed by experienced senior doctors can be presented to patients and their families through AR animation, so as to enhance the communication between doctors and patients. It can help the operator to clarify the risks and possibilities of complex surgery to patients.

The application prospects of AR animation technology are in pancreatic surgery education. With the continuous development of the software and hardware in AR technology, a surgical simulation training system combined with AR animation design for different parts of pancreatectomy can be established in the future. This will not only provide safe, realistic, practical, and repeatable surgical training methods for medical students, graduate students, and junior doctors, but also avoid adverse effects on patients and better achieve the minimally invasive requirements of modern surgery.

6   Conclusion

The combination of AR technology and animation makes animation more scientific and attractive. The combination of pancreatic surgery and AR animation has opened up a new approach for the development of medical education and clinical practice. Significant changes have occurred with respect to animation creation. With the rapid development and popular is ation of mobile digital media, the functions of mobile digital media devices are becoming increasingly excellent, providing a good, portable, and mobile platform for AR animation. With the development of AR animation in such a good environment, AR animation has broad application prospects in the field of medical education and clinical practice. Although this technology is still in the development stage and many challenges have yet to be overcome, the continuous development of AR animation technology is expected to make a major contribution to medical education and clinical practice in the future.

Acknowledgments. We would like to acknowledge the support of the Scientific Research Planning Project of Jilin Provincial Department of Education Project under grant JJKH20201230SZ.

References

1. Badash, I., Burtt, K., et al.: Innovations in surgery simulation: a review of past, current and future techniques. Ann. Transl. Med. **4**(23), 453 (2016)
2. Wu, H.K., Lee, S.W.Y., et al.: Current status, opportunities and challenges of augmented reality in education. Comput. Educ. **62**, 41–49 (2013)
3. Pessaux, P., Diana, M., et al.: Towards cybernetic surgery: robotic and augmented reality-assisted liver segmentectomy. Langenbeck's Arch. Surg. **400**(3), 381–385 (2015)
4. Yang, Y., Zhou, Z., et al.: Application of 3D visualization and 3D printing technology on ERCP for patients with hilar cholangiocarcinoma. Exp. Ther. medicine **15**(4), 3259–3264 (2018)
5. Lancet, T.: Violence against doctors: why China? Why now? What next. Lancet **383**(9922), 1013 (2014)
6. Wei, P., Feng, Z.Q., et al.: FEM-based guide wire simulation and interaction for a minimally invasive vascular surgery training system. In Proceeding of the 11th World Congress on Intelligent Control and Automation, vol. 6, pp. 964–969. IEEE (2014)
7. Watson, K., Wright, A., et al.: Can simulation replace part of clinical time? Two parallel randomised controlled trials. Med. Educ. **46**(7), pp. 657–667 (2012)
8. Bokken, L., Rethans, J.J., et al.: Instructiveness of real patients and simulated patients in undergraduate medical education: a randomized experiment. Acad. Med. **85**(1), 148–154 (2010)

Human Skeleton Control with the Face Expression Changed Synchronously

YueKun Jing[1], CongCong Wang[2], ZhiGeng Pan[1,3], ZhenWei Yao[1],
and MingMin Zhang[2,3]([✉])

[1] HangZhou Normal University, Hangzhou, China
[2] ZheJiang University, Hangzhou, China
zhangmm95@zju.edu.cn
[3] Guangzhou NINED Digital Technology Co., Ltd., Guangzhou, China

Abstract. In this paper, we proposed a method of real-time human motion detection and skeleton control based on the depth image. For the human body part, we use Kinect to detect the original three-dimensional coordinates of the human joint; then smooth the joint movement; calculate the amount of rotation based on the smoothed data. Finally, the FBX file is used to drive the virtual mannequin. For the human face part, we use 3D modeling software to modify the FBX model and replace the head of the human model. With the help of the Kinect device, we combined the idea of linear blend skinning (LBS) to make the human body action effect more realistic. The experimental results show that the motion capture system can better recover the 3D skeleton motion of the captured real human body. Compared with other motion capture system, the motion control time of the human body is greatly reduced, and the 3D mesh and skeleton can be obtained immediately according to the detected human body motion. The modified FBX model can apply specific expression effects.

Keywords: Skeletal animation · Motion capture · Virtual human · Kinect

1 Introduction

Active sensors were first used in human body tracking to obtain the motion information of human joints [1, 2]. The principle is to obtain the motion characteristics of different parts by using various sensors worn on the human body, and analyze the movement behavior according to specific algorithms. At present, to track human motion in the scene, cameras are widely used to obtain image sequences to extract human motion features and analyze motion recognition [3]. The most popular is the Kinect depth camera released by Microsoft which can combine 3D virtual reality, augmented reality and human-computer interaction. The RGB-D depth camera provides a lot of valuable data and greatly simplifies monocular posture reconstruction. This kind of human motion tracking based on image sequence mainly includes: extracting the complex human body from a complex environment, and obtaining human motion parameters (human body node trajectory information) through human motion tracking.

Z. Pan et al. (Eds.): Transactions on Edutainment XVI, LNCS 11782, pp. 154–162, 2020.
https://doi.org/10.1007/978-3-662-61510-2_15

The skeletal posture estimation of the 3D monocular camera has serious pathological problems. Some researchers have introduced the popular deep learning method, and achieved remarkable results. Movement tracking based on the neural network can identify more gestures and solve the occlusion problem in different degrees [4–6]. Mehta et al. [4] achieved a rendering rate of 30 frames per second by combining image algorithm, CNN and dynamic constraints. Pavlakos et al. [5] designed two neural networks by using probability and statistics model (SMPL [7]). The upper network can learn the posture parameters of the human body and the lower network can learn the shape parameters of the human body. With the help of SMPL model, Bogo et al. [6] proposed the idea of using capsules to approximate human body parts in order to reduce the amount of network computation, which solved the collision detection and occlusion problems of the model to some extent. Holden et al. [8] proposed a trajectory planning method and designed a network where the underlying parameters were converted into higher-order parameters, which could generate movements better. Tong et al. [9] proposed a data-driven method, which saved the data detected by the sensor into the BVH model file after processing, and used the model to demonstrate the movement effect of human body tracking.

There are many techniques for reconstructing a face model. It includes methods such as multi-view, orthogonal principle and probability statistics [10, 11]. Zhou et al. [13] used Radial Basis Function (RBF) interpolation to represent face deformation, used Local Binary Patterns (LBP) combined with local facial features, used support vector machine (SVM) as a classifier. This method is a sample to operate and has better deformation effect and higher recognition rate.

The contributions of this paper are as follows: (1) The FBX model is modified to satisfy the human body movement and the facial expression deformation; (2) human body model and face model fusion, which means that different human body models can be combined with different face technologies to generate better fusion effect.

2 System Framework

The whole system is divided into six parts, which are virtual human module, model modification module, human expression module, human tracking module based on Kinect depth map, 3D virtual human movement module and rendering module, respectively. The first two parts ensure that the model can be used for skeleton control and expression transformation. The third part determines the face deformation technology. The fourth part determines the human body tracking technology. The fifth part determines the skeleton control technology. And the last module determines the rendering technology. See Fig. 1.

For mannequin making, we hope to make a character that is suitable for use. 3D virtual software (MAKEHUMAN) can exports the 3D model file in different formats, which can be parsed by the ASSIMP library.

For model modification, we want the model to work for our technology. Unfortunately, mannequin file cannot be directly used, because the face parts exported by this software cannot be directly processed by expression module. We did some processing. Operation is as follows: the head of the original model is directly deleted using 3D modeling software; we manually merge the human body model and the deformed face model.

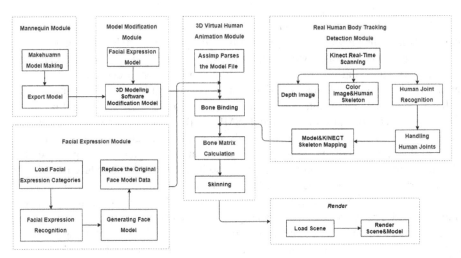

Fig. 1. System framework

The face model is made by other technologies and does not belong to the original model. It should be noted that the blending weights of the human model are totally deleted after removing the head of the model in Maya or 3dsmax, which should be reset manually.

Facial deformation has two functions of recognition and deformation. Therefore, the facial expression deformation with video stream as input can correct the position and direction of the face in real-time. Operation is as follows: global face features are fused with local features of eyes, nose, mouth and other images, and SVM classifier is used to realize face recognition.

The real-time human body tracking module detects human joint position in real-time based on Kinect depth map. According to the joint data obtained by Kinect, the joint transformation matrix and translation matrix can be calculated. Because the number of skeleton joints does not match, we also need to map the skeleton of FBX&Kinect [14].

3D animation module is the core of 3D virtual human motion control. This part requires data from all other modules. We use bone data parsed by the ASSIMP to match Kinect skeleton structure. The skeleton after the structure is processed can move correctly. After skeleton binding, rotation vector of each skeleton on the model should be calculated.

The rendering module uses the linear blend skinning (LBS) to calculate the final position of the vertices of the human mesh, and renders the final scene using Open Graphics Library.

3   Model Formulation

In this section, we first describe the skeleton control calculation and the model skinning method. Skeleton control is solved recursively using forward dynamics. The joint transformation matrix is calculated one by one according to the order of the parent joint to the sub-joint, so that the final pose transformation of the entire model is obtained, just like the hinge movement. In the rigid skinning algorithm, the motion of the vertices on the skin surface is affected by only one bone, so fracture occurs at the joints of the bones. In the linear blend skinning algorithm of flexible skinning, the vertices of the skin surface are affected by multiple bones. Therefore, this paper adopts the linear blend skinning algorithm, which is easy to use and memory saving, and has been widely used. Finally, we introduce the common part of facial expression module and 3D animation module, which can be combined with human action and expressions.

3.1 Skeleton Control

The character model used in our framework is a joint articulation consisting of only ball joints (3 DOF joints) and a 6 DOF root located at the pelvis. a skeleton $\mathbf{Q}$ of this character is described by $Q = (p^r, q^j), j = 1, \ldots, m$, where p^r is the root position and $q^j (j = 1, \ldots, m)$ describe the orientation of the joint, include the root. A motion P(t) is a function of time that relates each time value to a pose of the character $P(t) = Q_t$. The operation

$$P_t^j = Pos(Q_t, j) \tag{1}$$

is used to denote the computation of the position of the j-th joint of Q_t

As mentioned above, we mapped the skeletons of the Kinect and FBX models. The map function

$$Q = \text{Map}\left(Q^k, Q^f\right) \& k \neq f \tag{2}$$

explains the relation between the two skeletons. Q^k, Q^f are the Kinect and FBX skeletons respectively; k, j are the number of Kinect and FBX skeletons, respectively. We use the processed skeleton to drive character.

Rotations in R^3 can be described by axis angle that can explain the meaning of the joint orientation. For the calculation, the axis angle needs to be converted to a matrix using the R formula. The operation is as follows

$$U = (k, \theta) \& k = \begin{bmatrix} k_x \\ k_y \\ k_z \end{bmatrix} \tag{3}$$

$$R = \begin{bmatrix} cos\theta + (1 - cos\theta) \cdot k_x^2 & (1 - cos\theta) \cdot k_x k_y - sin\theta k_z & (1 - cos\theta) \cdot k_x k_z + sin\theta k_y \\ sin\theta k_z + (1 - cos\theta) \cdot k_x k_y & cos\theta + (1 - cos\theta) \cdot k_y^2 & (1 - cos\theta) \cdot k_y k_z - sin\theta k_y \\ (1 - cos\theta) \cdot k_x k_z - sin\theta k_y & (1 - cos\theta) \cdot k_y k_z + sin\theta k_y & cos\theta + (1 - cos\theta) \cdot k_z^2 \end{bmatrix} \tag{4}$$

where k is the position of rotation axis and θ is the angle of rotation axis. Exactly, R is our final goal.

The matrix obtained by R is only a representation in the local coordinates. we need to additionally calculate the joint matrix in the world coordinates. The world coordinates function P^w is determined by

$$P^w = P^p * P^g \tag{5}$$

$$P^g = T * R * S \tag{6}$$

where P^w, P^g are the joint position in the world coordinates and local coordinates, respectively and T, R, S represent the translation, rotation, and scaling matrices in the local coordinates, respectively.

3.2 Blending Model

Firstly, the transformation matrix of each joint was calculated according to the joint data collected by Kinect, and then the joint mapping of two different skeletons was conducted manually. Finally, the linear blend skinning algorithm was used to calculate the final grid vertex position of the FBX model. Given input parameters $\{\omega, P^w, Inv, V\}$, the blend model V' is determined by

$$V' = \sum_{i=1}^{n} \omega_i P_i^w Inv_i V \ \& \ \sum_{i=1}^{n} \omega_i = 1 \tag{7}$$

where i is the bone index and n is how many bones affect the movement of the skin vertex. V', V represents the position of the final transformed mesh vertex and the original world coordinates, respectively. ω_i represents the weight of the influence of the i-th joint on the skin vertices.

3.3 Connection of Human Movement and Expression

Makehuman's mannequins can be used for human movement, but can't control facial expressions changing. To process human motion and facial expressions simultaneously, we did frame alignment and model splicing. Model splicing can be used to transplant face models to show real-time face changes. The splicing model can display the facial expressions of different people after only one manual processing, and the "face skin changing" technology can be achieved by changing the facial texture in the later stage. Frame alignment solves the problem of inconsistency between facial expression changes and human body movements at different times.

Frame Alignment. Face technology generates different models by recognizing video facial features, while Kinect detects human skeleton to control human motion; The two technologies are not recorded in real-time at the same time, which leads to face recognition and human movement can't completely match. Given face video sequence F_t, a new mesh vertex model is determined by

$$V_t^{new} = Fa(V_{t1}', F_{t2}) \tag{8}$$

where t is the frame index and V_{t1}' is the character motion sequence frame.

Model Assembly. For the model splicing problem, it is easy to use software such as Maya or 3dsmax to adjust the shape of the model in real-time. We divide the V_t^{new} into two parts, V_{head}^{new} and V_{body}^{new}. We want

$$\left\{ V_{head}^{new}, V_{body}^{new} \right\} \rightarrow \left\{ V_{new\_head}^{new}, V_{body}^{new} \right\} \tag{9}$$

the old model head to be replaced. V_{head}^{new} is the original head model while $V_{new\_head}^{new}$ is the new head model generated by face technology.

4 Experimental Results

4.1 Preprocess

Though the skeleton jitters could be caused by the application performance due to both software and hardware, there are several internal possible reasons for skeleton joints jittering. In fact, Kinect's own smoothing method is still able to satisfy the common situation. Since the model requires a high level of motion, it also needs to consider real-time. For the more demanding real-time mannequins, the smoothing effect of Kinect is not enough. So we need to smooth the data. In this paper, mean filtering and median filtering are mainly adopted [14]. The joint data dimension is small and the filter core is small, so the entire filtering operation does not take up time.

Human motion is controlled by the skeleton. The universal human skeletons are CMU skeleton (31), Kinect skeleton (25) and SMPL skeleton (21) respectively. So not all joints require matrix calculations. Finally, we have selected 16 joints with rotation matrix operations. Note that the FBX bone is different from the Kinect bone, and their skeleton coordinates are just symmetric along the Y axis.

The original FBX model cannot be applied to the existing facial expression technology. We manually separate and fuse the face part of the model. The new model can not only achieve the original action effect, but also be compatible with the expression deformation effect of different technologies.

4.2 Results and Analysis

4.2.1 Human Action Effect

The experiment shows that the motion reconstruction under the monocular camera, has high realistic quality, which proves the effectiveness of the proposed method. The effect is shown in Fig. 2.

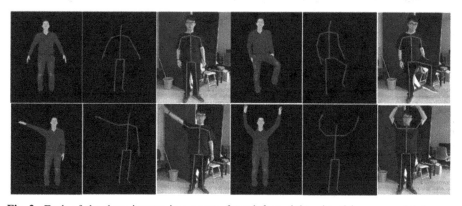

Fig. 2. Each of the three images is a group, from left to right: virtual human model, human skeleton, colour map detected by Kinect respectively. Here are four sets of actions, upper left: standard action, upper right: lifting leg, bottom left: lifting arm, bottom right: lifting arms.

Compared with human motion control based on neural network, Lassner et al. [6] and Pavlakos et al. [5] designed a more complex neural network by under probability and statistics model (SMPL [7]), which can better fit human posture and track human3D joints. However, these systems have a strong dependence on GPU, far from real-time control. It is not suitable for real-time requirements.

Compared with the skeleton movement detected based on Kinect, tong et al. [9] proposed a method to generate standard movement data in real-time. This method calculated the original data of human skeleton joints detected by Kinect, obtained joint transformation, and saved it as BVH file. By using BVH files to drive 3d skeleton motion, this design makes the system unable to be demonstrated in real-time and can't be automated. Figure 3 compares the two effects.

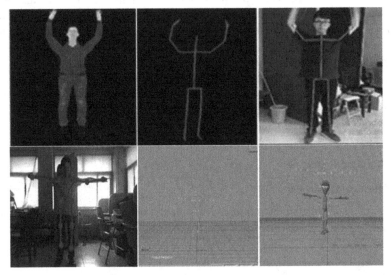

Fig. 3. The upper three are our demos, which is rendered in real-time and the bottom three are the demos of Tong et al., which uses other software to demonstrate.

4.2.2 Expression Fusion Effect

The fusion model shows the desired effect, and the action effect is consistent with Fig. 3. The most important thing is that the expression recognition rate is not reduced. The Fig. 4 shows the effect of the fusion of facial deformation and human action.

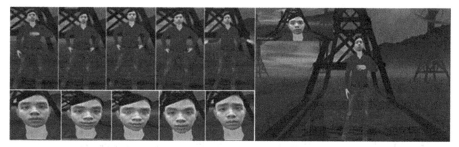

Fig. 4. The five human models on the left corresponds to the five facial expressions below, from left to right are nature-happy-enjoy-anger-sadness. On the right is the rendering of the whole human body movements and expressions. There are two cameras in the whole scene, one for observing Human body movement, the other for close-up of facial expressions.

5 Conclusion

This paper presents a method of human motion detection and skeleton control, and introduces the fusion of human motion and facial expression. For the face parts, we use radial basis function interpolation to realize the fusion of human motion and face model by manually adjusting the model. It satisfies the real-time human skeleton tracking and control on common equipment. When combined with the face model, the effect of the character is not much worse, but it takes longer and is still within the acceptable range. Combined with the human face model, the human body model can be changed into different faces according to different people. The character model does not need to be manually adjusted again, but only needs to cover the previous face texture of the model with the new face texture in advance. For human motion detection and tracking, we use Kinect to achieve the skeleton control. The advantage of doing this is to achieve real-time requirements, while achieving the basic character motion. According to the experimental results, the bone movement data collected by Kinect device can accurately control the virtual human movement. In the system design, we separate the whole project function into modules, so that the code can be transplanted to other motion capture systems.

Acknowledgment. We would like to acknowledge the support of the National Key Research and Development Project (Grant No. 2017YFB1002803) and the Guangzhou Innovation and Entrepreneurship Leading Team Project under grant CXLJTD-201609.

References

1. Yin, J., Yang, Q., Pan, J.J.: Sensor-based abnormal human-activity detection. IEEE Trans. Knowl. Data Eng. **20**(8), 1082–1090 (2008)
2. Moeslund, T.B., Hilton, A., KrÃijger, V.: A survey of advances in vision-based human motion capture and analysis. CVIU **104**(2–3), 90–126 (2006)
3. Wu, D., Pigou, L., Kindermans, P.J., et al.: Deep dynamic neural networks for multimodal gesture segmentation and recognition. IEEE Trans. Pattern Anal. Mach. Intell. **38**(8), 1583–1597 (2016)

4. Mehta, D., Sridhar, S., Sotnychenko, O., et al.: VNect: real-time 3D human pose estimation with a single RGB camera. ACM Trans. Graph. **36**(4), 44 (2017)

5. Pavlakos, G., Zhu, L., Zhou, X., et al.: Learning to estimate 3D human pose and shape from a single color image. In: The IEEE Conference on Computer Vision and Pattern Recognition (CVPR), pp. 459–468 (2018)

6. Bogo, F., Kanazawa, A., Lassner, C., Gehler, P., Romero, J., Black, Michael J.: Keep it SMPL: automatic estimation of 3D human pose and shape from a single image. In: Leibe, B., Matas, J., Sebe, N., Welling, M. (eds.) ECCV 2016. LNCS, vol. 9909, pp. 561–578. Springer, Cham (2016). https://doi.org/10.1007/978-3-319-46454-1_34

7. Loper, M., Mahmood, N., Romero, J., et al.: SMPL: a skinned multi-person linear model. ACM Trans. Graph. **34**(6), 248 (2015)

8. Holden, D., Saitoy, J., Komuraz, T.: A deep learning framework for character motion synthesis and editing. ACM Trans. Graph. (TOG) **35**(4) (2016)

9. Tong, X., Xu, P., Yan, X.: Research on skeleton animation motion data based on kinect. In: Proceedings of the 2012 Fifth International Symposium on Computational Intelligence and Design, ISCID 2012, vol. 2, pp. 347–350 (2012)

10. Meyer, G.P., Do, M.N.: Real-time 3D face modeling with a commodity depth camera. In: 2013 IEEE International Conference on Multimedia and Expo Workshops (ICMEW), pp. 1–4. IEEE (2013)

11. de Farias, M.M.C., Apolinário Jr., A.L., dos Santos, S.A.C.: Kinect fusion for faces: real-time 3D face tracking and modeling using a kinect camera for a markerless AR system. SBC J. Interact. Syst. **4**(2), 2–7 (2013)

12. Blanz, V., Vetter, T.: Face recognition based on fitting a 3D morphable model. IEEE Trans. Pattern Anal. Mach. Intell. **25**(9), 1063–1074 (2003)

13. Zhou, S.: The research and implementation of virtual human face modeling and 3D emotion interaction of facial expression. CNKI, ZJU (2019)

14. Wang, C.: Adaptive 3D virtual human modeling and animation system in VR scene. CNKI, ZJU (2018)

Context Construction and Virtual Human Based Intelligent Navigation in Virtual Experiment

Zhigeng Pan[1,3,4], Dan Yu[2], Yongheng Li[2], and Mingliang Cao[2(✉)]

[1] Guangdong Academy of Research on VR Industry, Foshan University, Foshan 528000, China
[2] Automation College, Foshan University, Foshan 528000, China
408177773@qq.com
[3] NINED Digital Technology Co., Ltd., Guangzhou 510000, China
[4] DMI Research Center, Hangzhou Normal University, Hangzhou 310012, China

Abstract. In view of the problems of high cost and many potential hazards in real middle school experiments, we construct a virtual experiment simulation situation with virtual human guidance in experiment operation, intelligent algorithm assistance in experiment failure, and systematic evaluation at the end of experiment. First of all, the experimental rule library is established, including: experimental process, trigger standard, coding standard and error content. Secondly, based on these libraries, the flow of system algorithm is designed. Finally, take the experiment of heating potassium permanganate to produce oxygen as an example. The flow design, scene design and virtual human design of the experiment are given. Realize intelligent assistance and interactive real-time guidance of the experimental process.

Keywords: Virtual experiment · Rule library · System algorithm · Virtual human

1 Introduction

Teaching oriented experiments are faced with complex natural and dangerous factors. With the higher requirements and difficulty of the experiment, the randomness of the results is more complex. It puts forward higher requirements for the comprehensive quality of students. With the development of virtual reality, especially the maturity of virtual reality hardware developed by HTC, Ocolus and other manufacturers, it provides a new idea for high-risk experiments. That is to use virtual reality technology as a means of display, high-simulation, full-scale hardware into the computer system. Considering the high degree of immersion and strong interaction in the operation of complex mechanical systems. The virtual human, one of the important VR interactive technologies, can be used to create different complex experiment scenarios [1–3]. The addition of the high-fidelity virtual human not only makes the experiment system more convincing, but also enhances the user's interest.

The design and integration process of virtual experiment is a complex and diverse task [4]. The University of Maryland began using virtual human in virtual teaching [5]. Adele [6], a teaching agent proposed by Johnson and Shawe of the University of Southern California (USE), is a successful application of virtual human. The most important

© Springer-Verlag GmbH Germany, part of Springer Nature 2020
Z. Pan et al. (Eds.): Transactions on Edutainment XVI, LNCS 11782, pp. 163–171, 2020.
https://doi.org/10.1007/978-3-662-61510-2_16

part of Adele's structure is agent and inference engine. It has basic educational functions: expressing knowledge, supervising students, providing feedback, probing into skillful problems, prompting and answering. Steve [7], a teaching agent developed by the Computer Department of the University of Southern California's Academy of Information Sciences, can be used to support students' learning process. Steve's architecture includes LAN-Da modules of perception, cognition and motion control. Avatak [8] is an intelligent virtual human system used for interactive skills training. It uses proxy technology to design a body double with knowledge and emotional expression. It consists of three parts: language processor, behavior engine and virtual human visualization engine. Jacob [9] project integrates various disciplines and technologies, including intelligent tutor system, virtual reality technology, intelligent agent technology, natural language processing technology, agent visualization, animation technology, etc. Whizlow [10], developed by North Carolina State University, has a virtual environment of H dimension and a lifelike agent, which provides learning guidance on computer architecture. Jack [11] is a software system developed in the human body modeling and simulation of Bingxifajiu University to control and simulate realistic characters in the H-dimensional environment. It provides collision detection, real-time object grabbing and real-time visualization functions. Zhao Huiqin of Beijing Normal University studied the emotional expression of virtual teachers and put forward the emotional model of virtual teachers [12]. In addition, some scholars have done some research on virtual human navigation [11, 13, 14].

Considering that middle school students have insufficient knowledge and limited experimental operation skills. Meanwhile teachers are unable to teach the student one by one. So there is a need to construct an experimental simulation scenario with virtual human guidance for experimental operation, intelligent algorithm assistance for experimental failure, and systematic evaluation at the end of the experiment. It is a challenging task for the system to detect and assist the experimenter in case of operation errors. In order to build the simulation context in the virtual experiment of middle school, focusing on the requirements and auxiliary support problems caused by the failure of the experiment, the experimental process library, trigger standard library, coding standard library and error content library are established. Based on the content library, the algorithm of the system is designed, thus realizing intelligent auxiliary and interactive real-time guidance of the experimental process.

2   System Design

2.1   Design Principle

The system design follows three principles:

(1) Realistic scene. Virtual experiment is a kind of image simulation of real experiment in virtual environment. In order to achieve the teaching effect of real experiment, virtual experiment not only needs to carry out highly realistic simulation of real experiment in scene design, but also needs to build a virtual teacher to show the teacher's actions and expressions.

(2) Algorithm intelligence. According to the experimenter's operation, the algorithm can track the current progress of the experiment, and can detect and judge the wrong operation.

(3) Real-time interaction. In order to make the experimenter to get personalized help, the virtual teacher should guide the students to operate the experiment and give real-time feedback on the operation.

2.2   Framework of System

The framework of system is shown in Fig. 1. This system is composed of experiment component, rule library, algorithm and experimental output. The user get feedback according to user operation, algorithm and rule library. Feedback includes effects of experimental results and guidance of the virtual teacher.

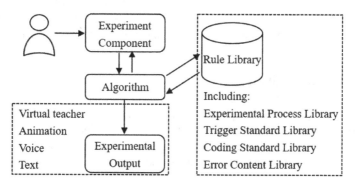

Fig. 1. Framework of system

2.3   Rule Library and System Algorithm

2.3.1   Rule Library

2.3.1.1 Experimental Process Library

The experimental process can be divided into three parts: operation object, operation constraint and experimental output.

(1) Operation objects can be divided into three types: movement, connection and disconnection. Movement refers to the designated movement of the experimental components. The connection means that two experimental components or combinations need to be connected, while the disconnection means that some components need to be separated from the combination of experimental components.

(2) Operational constraints can be divided into two types: spatial constraints and temporal constraints. Space constraint is a further requirement for connection operation, and time constraint is a requirement for the time interval between two operation steps.

(3) The experimental output includes experimental phenomena, component attribute changes and new moving trigger standards.

2.3.1.2 Trigger Standard Library

Trigger standard library refers to the mathematical model of operation standard. The criteria for component triggering can be divided into object triggering and space-time constraint. The object trigger standard refers to the mathematical model of components' movement, connection and disconnection. The space-time constraint standard refers to the mathematical model of the components triggered in space or time.

2.3.1.3 Coding Standard Library

The coding standard library refers to the standard coding state of component combinations and attributes corresponding to experimental operations. The coding rules of components are: according to the different names and attributes of components, they are respectively represented by 2^n, $n = 0, 1, 2...$. The operation rules of components are: connected equipment, code addition, equipment with changed attributes, and code conversion.

2.3.1.4 Error Content Library

The error content library include: sequence number (primary key), error type, error operation description, abnormal phenomena, and relevant knowledge. There are three types of errors: object, space and time. Error operation description refers to the specific description of the error operation. Abnormal phenomenon refers to the consequences of wrong operation. Relevant knowledge refers to the experimental knowledge involved in this operation.

2.3.2 System Algorithm

After receiving the trigger, the system first judges whether the trigger result is consistent with the standard code in the coding standard library, and then judges whether it meets the conditions of space-time constraint. If an operation error is detected, the system will access the knowledge of the error content library. If the operation is correct, the system will generate corresponding feedback according to the experimental process library. The algorithm flow is shown in Fig. 2.

2.4 Intelligent Navigation Virtual Human Application

Compared with traditional virtual experiments, intelligent navigation virtual human embodies the combination of interest and knowledge, and provides a reference for exploring the deep integration of the intelligent behavior mode. The application is equipped with an intelligent navigation virtual human. Providing voice messages and using it's animations to display a powerful visual performance will help improve understanding and motivation levels. The intelligent virtual human has several features: (1) Interacting with virtual human enables users to gain real experimental experience based on life experience; (2) The virtual human helps the user understand the experimental process

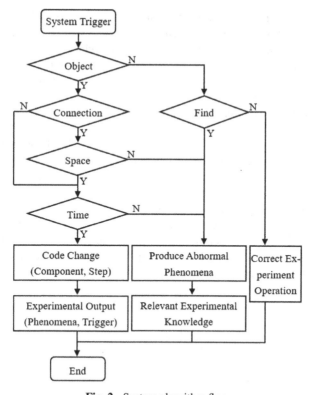

Fig. 2. System algorithm flow

and guide the experimental steps; (3) The virtual human can link knowledge databases, store user behavior data, and provide feedback.

3   Experimental Result

Take the experiment of heating potassium permanganate and preparing oxygen by drainage method as an example.

3.1   Process Design

As shown in Fig. 3, the process design is divided into seven parts, as follows:

(1) Draw a flow chart of the experiment, including experimental operation and experimental phenomena.
(2) The experimental process library includes: which object the experimental operation belongs to, whether space or time constraints are needed, whether phenomena occur, whether attributes change, and whether triggers are newly created. For example, when fixing the test tube, the operating object belongs to the connection between the iron clamp and the test tube, and needs to be fixed at about 1/3 of the test tube

opening. There is no phenomenon or attribute change, but there is a moving trigger with a new test tube opening slightly inclined downward.

(3) The coding standard library refers to the standard coding state of component combinations and attributes corresponding to experimental operations. For example, the code of the catheter is 1 and the code of the rubber stopper is 4. When the catheter is inserted into the rubber stopper, the code is changed from 1 and 4 to 5.

(4) Error content library refers to the description, abnormal phenomena and related knowledge corresponding to error experiment operation. For example, if the test tube is not preheated, it is heated by concentrated fire. The abnormal phenomenon is that the test tube bursts, and the related knowledge is that the test tube is heated unevenly.

(5) According to the requirements of the experiment on scenes, equipment and phenomena, the components needed for the experiment are designed.

(6) Trigger standard library refers to the mathematical model of operation standard. For example, the operation of placing the bottle cap on the experimental table. The standard distance between the bottle cap and the test bench, and the standard angle of upside-down and upright position.

(7) Design the algorithm of the system.

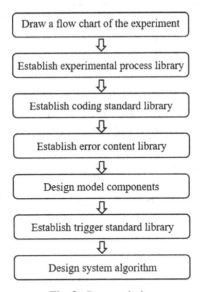

Fig. 3. Process design

3.2 Construction of the Virtual Experiment

We developed the basic interactive functions and key tasks of the virtual experiment based on the experience of existing virtual experiment [15]. In the virtual experiment, a basic interactive task is to control the visual angle through the HTC Vive helmet, and to manipulate the HTC handle to simulate the left and right hands, then use the HTC handle to point and reach the object and click on the handle button to interact. The user touches the selected instrument and picks it up. The data of out virtual experiment environment scene model can accurately grasp the entire modeling size and the positioning of each building location through the architectural CAD drawing data. The laboratory model is made using polygon modeling method of 3DSMax to transform the geometry into an editable polygon. The experimental equipment was made according to the size of the real experimental model. Then the model is imported into Unity3D to simulate in Fig. 4.

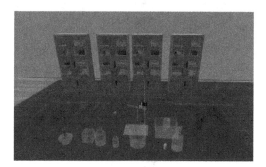

Fig. 4. Virtual experiment environment

3.3 Virtual Human Design

Below is an application showcase for intelligent navigation virtual human: interaction, correctitude and evaluation in Fig. 5. The intelligent navigation virtual human performs the following guidance:

(1) Interaction – as shown in Fig. 5(a), the user clicks on the button on the right hand side of the virtual human by HTC Vive.
(2) Correctitude – as shown in Fig. 5(b), if the user does not place the cap on the bench, the avatar will give a feedback note: "The cap is not standing on the table and will contaminate the test rig".
(3) Evaluation – in Fig. 5(c), the virtual human evaluates based on users' experimental process and gives a rating.

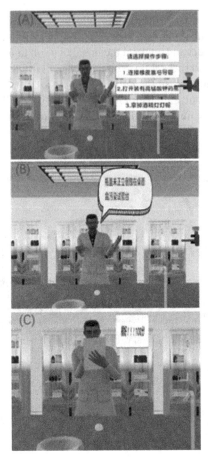

Fig. 5. Intelligent virtual human display (a) Interaction: experimental steps operation guidance (b) Correctitude: text box plus action indicates error (c) Evaluation: record every steps and evaluate

4   Conclusion

In this paper, the real context of the middle school experiment is simulated. According to the context requirements of real experiments, the rule base and algorithm are established. Take heating potassium permanganate to produce oxygen as an example. The flow design, scene design and virtual human design of the experiment are given. The project objectives of virtual human guidance for experimental operation, intelligent algorithm assistance for experimental failure, and systematic evaluation for the end of the experiment were completed.

At present, there are still many deficiencies in the simulation scenario we have built. The specific manifestations are as follows: 1. The experiment can only carry out a small number of parallel operations, and lacks the exploratory and open nature of the experimental operations. 2. The evaluation index is not comprehensive enough and the result lacks individuality. Therefore, we will further improve the openness of experimental operation and customization of experimental evaluation.

Acknowledgement. The author thanks the reviewers for their constructive and thoughtful comments, as well as teacher Cao Mingliang's guidance and Li Mingtang's help. This research was supported by the National Key Research and Development Program of China under grant no. 2018YFB1004902, and Key Scientific and Technological Innovation of Hangzhou under grant no. 20182014B02. Also, we acknowledge the support of the Guangzhou Innovation and Entrepreneurship Leading Team Project under grant CXLJTD-201609.

References

1. Ahn, S.J., Fox, J.: Immersive virtual environments, avatars, and agents for health. In: Encyclopedia of Health and Risk Message Design and Processing (2017)
2. Machidon, O.M., Duguleana, M., Carrozzino, M.: Virtual humans in cultural heritage ICT applications: a review. J. Cult. Herlt. **33**, 249–260 (2018)
3. Triandafilou, K.M., Tsoupikova, D., Barry, A.J., et al.: Development of a 3D, networked multi-user virtual reality environment for home therapy after stroke. J. Neuroeng. Rehabil. **15**(1), 1–13 (2018)
4. Slater, M., Antley, A., Davison, A.: A virtual reprise of the Stanley Milgram obedience experiments. PLoS ONE **1**(1), 39 (2006)
5. Gillespie, R., O'Modhrain, M., Tang, P., Zaretzky, D., Pham, C.: The virtual teacher. In: Proceedings of the ASME Dynamic Systems and Control Division, vol. 64, pp. 171–178 (1998)
6. Johnson, W.L., Shaw, E.: Using agents to overcome deficiencies in web-based courseware. In: Proceedings of the Workshop "Intelligent Educational Systems on the World Wide Web", 8th World Conference of the AIED Society, IV-2 (1997)
7. Rickel, J., Johnson, W.L.: Animated agents for procedural training in virtual reality: perception, cognition, and motor control. Artif. Intell. **13**, 343–382 (1999)
8. Guinn, C., Hubal, R.: Extracting emotional information from the text of spoken dialog. In: Proceedings of the 9th International Conference on User Modeling (2003)
9. Evers, M., Nijholt, A.: Jacob - an animated instruction agent in virtual reality. In: Tan, T., Shi, Y., Gao, W. (eds.) ICMI 2000. LNCS, vol. 1948, pp. 526–533. Springer, Heidelberg (2000). https://doi.org/10.1007/3-540-40063-X_69
10. Lester, J.C., Zettlemoyer, L.S., Gergoire, J.P., et al.: Explanatory life like avatars: performing user - centered tasks in 3D learning environments. In: Proceedings of the Third International Conference on Autonomous Agents, pp. 24–31 (1999)
11. Hu, M.W.: A high-fidelity three-dimensional simulation method for evaluating passenger flow organization and facility layout at metro station. Simul.-T. Soc. Mod. Sim. **93**(10), 841–851 (2017)
12. Zhao, H., et al.: 3D virtual teacher emotional expression based on body language. Comput. Eng. **37**(23), 159–164 (2011)
13. Seo, S., Kim, E., Mundy, P.: Joint attention virtual classroom: a preliminary study. Psycgiat. Invest. **16**(4), 292 (2019)
14. Rickel, J., Johnson, W.L.: Animated agents for procedural training in virtual reality: perception, cognition, and motor control. Appl. Artif. Intell. **13**(4–5), 343–382 (1999)
15. Makransky, G., Terkildsen, T.S., Mayer, R.E.: Adding immersive virtual reality to a science lab simulation causes more presence but less learning. Learn. A. Instr. **60**, 225–236 (2017)

CV and AI

Using CNN and Channel Attention Mechanism to Identify Driver's Distracted Behavior

Lu Ye[1,2(✉)], Cheng Chen[2], Mingwei Wu[1,2], Samuel Nwobodo[1], Annor Arnold Antwi[1], Chido Natasha Muponda[1], Koi David Ernest[1], and Rugamba Sadam Vedaste[1]

[1] School of Information and Electronic Engineering, Zhejiang University of Science and Technology, Hangzhou 310023, Zhejiang, China
yelue@zust.edu.cn
[2] School of Mechanical and Energy Engineering, Zhejiang University of Science and Technology, Hangzhou 310023, Zhejiang, China
Chen_Cheng066@163.com

Abstract. The driver's distracted attention will cause a huge safety hazard to the traffic. In different types of distraction, it is illegal to make phone calls and smoke while driving, which will be fined in China. In order to solve this problem, a method of driver's distracted behavior detection based on channel attention convolution neural network is proposed. SE module is added to the Xception network, which can distinguish the importance of different feature channels and enhance the expression ability of the network. SE module mainly assigns different weights to features to enhance more important features and suppress less influential features. The experiment uses Xception and SE-Xception for comparison. The experimental results show that the accuracy of SE-Xception is 92.60%, which has a good performance for the distracted driving behavior detection of drivers.

Keywords: Squeeze-and-Excitation · Channel attention · Convolution neural network · Distracted driving

1 Introduction

The development of social economy cannot be separated from the development of traffic. Traffic safety has become a serious problem that attracts worldwide attention, and the state and behavior of drivers are the main reasons that affect traffic safety. In the process of driving, drivers will inevitably conduct behaviors other than driving, which will lead to inattention while driving, thus causing hidden dangers of traffic safety. Some studies have shown that the probability of accidents increases 3 times and 4 times respectively when talking on the phone or sending short messages while driving [1]. In China, if driver is calling while driving will result in a fine of 100 yuan and a penalty of 2 points off your driver's license. If Drivers smoke while driving, they will be fined 50 yuan. Therefore, it is necessary to carry out relevant research on illegal driving behavior to reduce its potential traffic hazards.

© Springer-Verlag GmbH Germany, part of Springer Nature 2020
Z. Pan et al. (Eds.): Transactions on Edutainment XVI, LNCS 11782, pp. 175–183, 2020.
https://doi.org/10.1007/978-3-662-61510-2_17

The research on driver's behavior has attracted the attention of researchers all over the world. Related research can be roughly divided into two types: using traditional machine learning algorithm and deep learning algorithm. For the traditional machine learning algorithms, most of these researches design specific features by themselves and use traditional algorithms to distinguish drivers' distraction behavior based on these features. Kutila [2] obtained the data of the driver's vision, head and lane keeping differences. On this basis, support vector machine (SVM) method was used to detect and classify the driver's distraction while driving in real vehicles, and the recognition accuracy was between 65% and 80%. Sahayadhas [3] and their team proposed a detection method of driver's state based on physiological information. First, the driver was measured by electrocardiogram and electromyography to obtain physiological information, and then the features contained in the information were analyzed by K-neighborhood and linear discrimination methods. In the experiments of Ebadi [4], the researchers asked 24 drivers to simulate driving situations in 18 different scenes twice on a driving simulator. And researchers will monitor the eyes of the drivers at the same time. During the first simulation, hands-free handset tasks will be performed, while during the second, there will be no other influencing factors. The experimental results show that when the driver is driving on a hands-free mobile phone, the scanning ratio of the eyeball to the surrounding scene is significantly reduced, and the potential danger in the driving environment is less likely to be found. In the above method, some features are difficult to obtain and require some hardware equipment with higher cost.

At present, with the rise of deep learning methods, many researchers also try to apply this method to driver behavior detection. Most of these methods are based on computer vision, which only needs to obtain the driver's image information and automatically extract image features through deep learning algorithm. Le [5] tried to use the multi-scale fast RCNN method to detect whether the driver uses a mobile phone or how many hands are placed on the steering wheel, based on the information to detect the driver's state. The research achieved a good result. Xing et al. [6] used Kinect to collect the driver's behavior depth image information and combined it with joint position information as data. Then used random forest and maximum information coefficient method to obtain the importance of different features, and finally used feedforward neural network to identify the driver's different behaviors. Masood and Celaya-Padilla et al. [7, 8] both use convolution neural network to detect driver distraction behavior. The former defines one normal driving state and nine distraction behaviors, and the latter uses wide-angle camera installed in the car to acquire data to identify driver short message behavior and normal driving. Hu et al. [9] proposed a lightweight semi-cascade network structure, which extracts morphological features of human faces and hands to judge the driver's behavior.

The traditional method has the problems of difficult feature extraction, high cost, and certain limitations in recognition performance. Therefore, the deep learning method has great advantages. In this paper, SE module is added to the convolution layer of the feature extraction network based on the Xception network to further improve the expression ability of the network. This paper proposes a convolution neural network algorithm based on channel attention mechanism to identify driver distraction behavior.

2 Convolution Neural Networks

Convolutional neural network algorithm has excellent performance in the field of computer vision. The concept of convolution neural network (CNN) [10] was proposed by LeCun. The appearance of this network architecture has made breakthrough progress in the field of image processing.

2.1 Xception

The Inception structure is proposed in GoogLeNet [11], which mainly improves the traditional convolution layer to use convolution kernels of different sizes in the same convolution layer. It innovatively decompose a convolution kernel's process of mapping cross-channel correlation and spatial correlation into a series of independent operations. On this basis, there have been improved InceptionV2 [12], InceptionV3 [13], InceptionV4 [14]. What is more, Xception [15] uses depthwise separable convolution based on InceptionV3. The name of Xception module stands for "Extreme Inception", which firstly uses 1×1 convolution to map cross-channel correlation, and then has an independent spatial convolution on each output channel of 1×1 convolution to map spatial correlation. At the same time, a mechanism similar to residual connection is also added to Xception to speed up the convergence of the network. The Xception module is shown in Fig. 1.

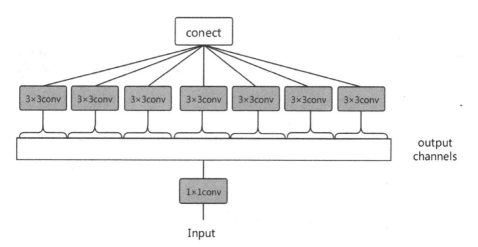

Fig. 1. Extream Inception module

2.2 SE Module

Squeeze-and-exclusion networks (SENet) [16] was proposed by researchers in Momenta, an autonomous driving company. SENet won the championship in the 2017 ILSVR competition. Attention and gating mechanism have always been an important research content for neuroscientists, SE module is a lightweight module applying this idea, which can be embedded into convolutional neural network. Through this mechanism, we can learn to use global information to selectively emphasize better features and suppress less useful features. The structure of the module is shown in Fig. 2.

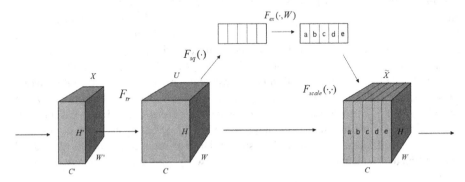

Fig. 2. Squeeze-and-Excitation module

Assume that the input of neural network is $X \in R^{W' \times H' \times C'}$, Generating a feature map $U \in R^{W \times H \times C}$ by convolution operation. If v_c represents the convolution kernel of the convolution layer, u_c represents the receptive field of the feature map at channel C. This step can be expressed by Eq. (1):

$$u_c = v_c * X = \sum_{s=1}^{c'} v_c^s * X^s \tag{1}$$

After that, using the global average pooled image layer to reduce the dimension of the image, and each two-dimensional feature channel is changed into a real number. This step is shown in Eq. (2):

$$Z_c = F_{sq}(u_c) = \frac{1}{W \times H} \sum_{i=1}^{W} \sum_{j=1}^{H} u_c(i, j) \tag{2}$$

After obtaining the channel information, in order to reduce the amount of calculation when processing data, the bottle structure of "full connection layer 1 - Relu activation function - full connection layer 2" is used. Finally, Sigmoid function is used to assign channels different weights, and this step can be expressed by Eq. (3):

$$s = F_{ex}(Z, W) = \sigma(W_2 \delta(W_1 Z)) \tag{3}$$

In this formula, z represents the operation result of the previous step, W_1 and W_2 respectively represent the parameters of the two fully connected layers. $\delta(\cdot)$ and $\sigma(\cdot)$

represent Relu function and Sigmoid function. The result obtained in this step is the weight value corresponding to different channels, set to $\omega = a, b, c, d, e$.

Finally, the result of the output is re-calibrated on the original image in a weighted manner. If represents weight, this step can be represented by Eq. (4):

$$\tilde{X}_c = F_{scale}(u_c, s_c) = s_c \cdot u_c \tag{4}$$

In this way, the network can strengthen the effective channel and improve the characterization ability of the feature map.

Based on the original Xception network architecture, an SE module is added at the end of all convolutional layers. The addition of this module did not change the structure and output of the feature extraction part of the original network. After the addition of this branch module, the sensitivity of the network to different feature channels of the image was enhanced, and effective features were further extracted by channel correlation. The specific network structure is shown in Fig. 3.

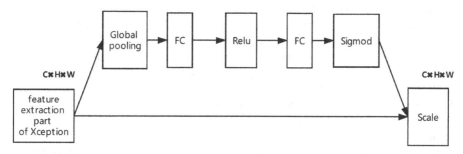

Fig. 3. Structure of SE-Xception

3   Experiment

3.1   Dataset

For the Chinese driver's illegal behavior during driving, the experiment uses the Chinese driver's illegal behavior data set established by ourselves. The data set was collected by Kinect equipment. The number of volunteers participating was 6 and the shooting place was the driving position of the test car. During shooting, the volunteers need to completed the prescribed different driving behaviors in sequence, and saved the data at the frame rate of 10 frames per second.

The collected image size is 640×480, and every image was assigned a label to describe its content. The data set mainly includes 3 different behaviors during driving, i.e. safe driving, calling while driving, smoking while driving, totaling 13416 images. The images of 5 drivers were selected as training sets to train the neural network model, and the images of the remaining 1 driver were selected as test sets. The image and the corresponding label are shown in Fig. 4.

(a) Safe driving

(b) Calling while driving

(c) Smoking while driving

Fig. 4. Image in dataset

3.2 Experimental Results and Analysis

The experimental operating environment is win10 64-bit operating system, GPU is GTX1060 6G, 16G of computer memory. The simulation is based on python, and the network model is built using keras and tensorflow frameworks.

In the experiment, the pre-trained Xception network was used to initialize the weights of feature extraction networks. The optimizer was defined as SGD, the momentum of SGD was 0.9, the learning rate was initialized to 1×10^{-3}, and the attenuation rate was 1×10^{-6}. The loss function uses the categorical cross-entropy function and takes the value of the loss function and the global accuracy as model evaluation. Resize the image to $299 \times 299 \times 3$ before inputting the data into the model and compile the model. Input the image and label into the model to train the weights of the CNN model. The number of training samples for each batch is 12 and 50 epochs are trained. If the loss function of the test set exceeds 10 consecutive epochs and does not drop, the training will be stopped in advance.

Save the model with the optimal loss function on the test set, record the accuracy of the model identification results, and then input the test set to the model and obtaining the corresponding confusion matrix of identification results to observe the identification effect of each behavior category.

In the experiment, Xception network and SE-Xception network added with SE module were used for comparative experiments, and the identification results were compared. The experimental results are shown in Tables 1 and 2.

Table 1. Result of Xception model

Label	Total samples	Correct predictions	Incorrect predictions	Accuracy (%)
Safe driving	657	495	162	75.34
Calling while driving	767	761	6	99.21
Smoking while driving	780	746	34	95.64
Overall_accuracy	2204	2002	202	90.83

Table 2. Result of SE-Xception model

Label	Total samples	Correct predictions	Incorrect predictions	Accuracy (%)
Safe driving	657	560	97	85.23
Calling while driving	767	754	13	98.31
Smoking while driving	780	727	53	93.21
Overall_accuracy	2204	2041	163	92.60

The experimental results show that the convolution neural network model has good performance for Chinese driver's illegal behavior recognition, the overall accuracy reaching more than 90%. From the confusion matrix of the two in Fig. 5, it can be seen that the two models have better recognition performance for calling while driving and smoking, which are both above 90%. And it's not much different. But at the same time safe driving behavior is easily confused with calling while driving. And the recognition accuracy of SE-Xception classifier added with SE module for safe driving is 85.23%, while the recognition accuracy of Xception model classifier for safe driving is only 75.34%. Generally speaking, the overall accuracy of SE-Xception is higher than that of Xception. The main reason is that after the feature extraction network, SE module gives different weights to feature channels to make the model selectively emphasize important features, suppress invalid features, and can explicitly model the interdependence between convolution feature channels to improve the representation capability of the network.

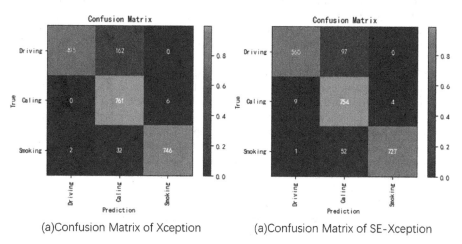

(a)Confusion Matrix of Xception   (a)Confusion Matrix of SE-Xception

Fig. 5. Confusion matrix

4 Conclusion

In conclusion, this study proposes a good performance detection method for Chinese driver's illegal behavior, which is verified by experiments on self-made data sets. In this method, SE structure is added after the feature extraction part of convolutional neural network, which makes the network consider the importance of different feature channels. The channel-based attention mechanism enhances the feature expression ability of the network and enables the neural network classifier to have better recognition performance. In addition, there is still some work for improvement. Such as a complete driver's illegal behavior warning system can be developed based on this method, the driver's real-time state detection can be carried out through a camera fixed on the upper right of the driver. The system intercepts the video frames of video from camera at certain intervals, then judges the current driving state of the driver and records the result. If the driver's current state and the previous state records for several consecutive times (threshold) are all in distraction state, the driver can be warned and reminded using the corresponding equipment.

References

1. Choidhary, P., Velaga, N.R.: Mobile phone use during driving: effects on speed and effectiveness of driver compensatory behavior. Accid. Anal. Prev. **106**, 370–378 (2017)
2. Kutila, M.H., Jokela, M., Mäkinen, T., Viitanen, J., Markkula, G., Victor, T.W.: Driver cognitive distraction detection: feature estimation and implementation. Proc. Inst. Mech. Eng. Part D: J. Automob. Eng. **221**, 1027–1040 (2007)
3. Sahayadhas, A., Sundaraj, K., Murugappan, M., Palaniappanb, R.: A physiological measures-based method for detecting inattention in drivers using machine learning approach. Biocybern. Biomed. Eng. **35**, 198–205 (2015)
4. Ebadi, Y., Fisher, D.L., Roberts, S.C.: Impact of cognitive distractions on drivers' hazard anticipation behavior in complex scenarios. Transp. Res. Rec.: J. Transp. Res. Board **2673**, 440–451 (2019)
5. Le, T.H.N., Zheng, Y., Zhu, C., Luu, K., Savvides, M.: Multiple scale faster-RCNN approach to driver's cell-phone usage and hands on steering wheel detection. In: The IEEE Conference on Computer Vision and Pattern Recognition Workshops (CVPRW), Las Vegas (2016)
6. Xing, Y., et al.: Identification and analysis of driver postures for in-vehicle driving activities and secondary tasks recognition. IEEE Trans. Comput. Soc. Syst. **5**, 95–108 (2018)
7. Masood, S., Rai, A., Aggarwal, A., Doja, M.N., Ahmad, M.: Detecting distraction of drivers using convolutional neural network. Pattern Recognit. Lett. (2018)
8. Celaya-Padilla, J.M., et al.: Texting & driving detection using deep convolutional neural networks. Appl. Sci. **9**, 2962 (2019)
9. Hu, J., Liu, W., Kang, J., Yang, W., Zhao, H.: Semi-cascade network for driver's distraction recognition. Proc. Inst. Mech. Eng. Part D: J. Automob. Eng. **233**, 2323–2332 (2019)
10. Lecun, Y., Bottou, L., Bengio, Y., Haffner, P.: Gradient-based learning applied to document recognition. Proc. IEEE **86**, 2278–2324 (1998)
11. Szegedy, C., et al.: Going deeper with convolutions. In: The IEEE Conference on Computer Vision and Pattern Recognition (CVPR), Boston (2015)
12. Ioffe, S., Szegedy, C.: Batch normalization: accelerating deep network training by reducing internal covariate shift. Proc. Mach. Learn. Res. **37**, 448–456 (2015)

13. Szegedy, C., Vanhoucke, V., Ioffe, S., Shlens, J., Wojna, Z.: Rethinking the inception architecture for computer vision. In: The IEEE Conference on Computer Vision and Pattern Recognition (CVPR), Las Vegas (2016)
14. Szegedy, C., Ioffe, S., Vanhoucke, V.: Inception-v4, Inception-ResNet and the impact of residual connections on learning. In: The Thirty-First AAAI Conference on Artificial Intelligence, San Francisco (2016)
15. Chollet, F.: Xception: deep learning with depthwise separable convolutions. In: The IEEE Conference on Computer Vision and Pattern Recognition (CVPR), Honolulu (2017)
16. Hu, J., Shen, L., Sun, G.: Squeeze-and-Excitation networks. In: The IEEE Conference on Computer Vision and Pattern Recognition (CVPR), Salt Lake City (2018)

A Method of Human Motion Feature Extraction and Recognition Based on Motion Capture Device

Jinhong Li, Huaming Gao, Fengquan Zhang$^{(\boxtimes)}$, and Mingwu Zheng

School of Information Technology, North China University of Technology, Beijing, China
fqzhang@ncut.edu.cn

Abstract. In recent years, motion capture devices have been widely used in 3D film special effects, animation generation, digital media and other virtual reality fields. The purpose of this paper is to achieve data acquisition, motion feature extraction and recognition of human motion by using motion capture device. First, we use the OptiTrack motion capture device and related data preprocessing methods to collect motion data, and realize the preview of joint point offset data. Then, a human motion feature representation method which combines the collected data with the key frame is designed. For the selection of key frames, we improve the frame subtraction algorithm by adding the second derivative calculation of reconstruction error to achieve the number of key frames automatic determination. In addition, in order to solve the problem of huge amount calculations and error existing in discretization of observation state, we use high-dimensional gaussian function to fit human motion data, and finally apply the method of Gaussian-Mixture Hidden Markov Model (GMM-HMM) for motion recognition. Experiments show that the method has achieved remarkable performances in human motion extraction and recognition.

Keywords: GMM · HMM · Motion recognition · Feature extraction · OptiTrack

1 Introduction

With the development of computer vision, human motion recognition has become a hot research topic. It is not only widely used in intelligent monitoring, video retrieval, human-computer interaction and other aspects, but also involves a number of key technologies such as image processing, behavior representation, data processing and so on, which is a challenging interdisciplinary project [1].

Many researchers have explored methods of human behavior recognition. Yang et al. tried to analyze workers' behaviors in video monitoring using vision-based methods, so as to facilitate construction management, and collected video data set [2]. Liu et al. proposed a method of using Microsoft Kinect sensor, K-means clustering and Hidden Markov Models (HMM) to identify human behavior, and achieved 91.4% accuracy [3]. When Núñez studied, they used combination of a Convolutional Neural Network (CNN) and a Long Short-Term Memory (LSTM) recurrent network for skeleton-based human

© Springer-Verlag GmbH Germany, part of Springer Nature 2020
Z. Pan et al. (Eds.): Transactions on Edutainment XVI, LNCS 11782, pp. 184–195, 2020.
https://doi.org/10.1007/978-3-662-61510-2_18

activity and hand gesture recognition [4]. Min et al. proposed a human behavior recognition method based on sparse representation, which implicitly splits actions and context information in test video segments through dictionary construction, thereby improving the validity of classification [5]. Liang et al. used a framework for human behavior analysis using variable-length markov models (VLMMs), which combined the advantages of the excellent learning function of a VLMM and the fault-tolerant recognition ability of an HMM [6]. Li implemented a human motion recognition method based on fuzzy support vector machine, which employed the membership function to solve the unclassifiable areas in the traditional SVMs' two-class problems extending to the multi-class problems [7]. And Hussein et al. proposed a method to extract 3D skeleton sequences from depth data, which combined covariance descriptors with existing classification algorithms [8]. Baccouche et al. created an automatic depth model to classify human behaviors without using any prior knowledge [9]. Subramanian et al. proposed a sequential Meta-Cognitive learning algorithm for Neuro-Fuzzy Inference System (McFIS) to efficiently recognize human actions from video sequence [10].

Based on these previous research, we use the Gaussian-Mixture Hidden Markov Model (GMM-HMM) to identify human body motion and finally realize the design of a stage interaction system in this paper. In order to meet real-time requirements, offline training and online identification are conducted. This method can well improve the accuracy of human motion recognition.

2 Data Acquisition Based on OptiTrack Device

In Motive, the supporting software of OptiTrack equipment [11], human skeleton can be created to acquire data. As we know that the human skeleton is represented by the human joints, so we need to mark the key joint points of the human body when using the OptiTrack to capture the movement data of the characters. In this paper, we used 20 joint points as shown in Fig. 1.

In addition, the data acquired by OptiTrack has the defects of data fluctuation and data loss, which brings negative impact on human motion data preview, motion data editing, human motion recognition and human animation reconstruction. Therefore, the locally weighted linear regression algorithm [12] is used to estimate the predicted values so as to fill the vacancy data. Based on the linear regression algorithm, the locally weighted linear regression algorithm adds weight values. To assign a smaller weight to the data farther away from the predicted value, the influence of these longer distance data on the parameter training will be eliminated, and the data near the predicted data will receive more attention. The parameters of the locally weighted linear regression are updated as:

$$\theta_j^k := \theta_j^k + \alpha \sum_{i=1}^{i=n} W(t_i)(y_i^j - h_\theta(t_i))t_i^k \tag{1}$$

where α represents the learning rate and n represents the number of frames in a single sample. When updating the parameters, all the data are calculated together, and the data of each frame can be regarded as having the same weight. The weight matrix $W(t_i)$ is given by:

$$W(t_i) = e^{-\frac{(t-t_i)}{2\tau^2}} \qquad (2)$$

where t represents the time frame to be predicted, t_i represents the time frame currently calculated. τ is the wavelength parameter, and the larger the value of τ, the more susceptible the data to be predicted is to the longer distance data. In addition, the graph of the function is similar to the probability density function, which is a bell curve.

Although the algorithm is not directly related to the Gauss distribution function, we find it achieves better prediction results when the wavelength is set to a standard deviation, compared to that when it is set to 1. After many experimental tests, the learning rate is finally set as the reciprocal of the number of frames multiplied by one percent.

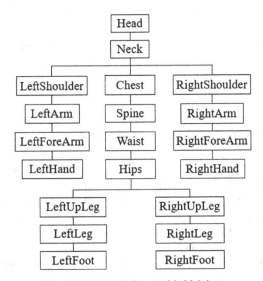

Fig. 1. The 3D skeleton with 20 joints.

3   Feature Extraction

3.1   Extracting Boundary Frames of Human Motion Data

For human motion data sequences, the boundaries of each dimension of data can well express the characteristics of human motion. At the same time, for data with a large number of frames, extracting the boundary frame sequence [13] can effectively reduce the amount of data. To extract the boundary frame of human motion data, first we need to find the maximum and minimum values of each dimension in the skeleton data; then we store the frame number where the maximum and minimum values are located, and the frame number is only saved once; finally, to ensure the obtained boundary frames still retain the original sequence nature of the motion data, the saved frame numbers need to be sorted.

Each motion data sample is defined as:

$$Y = \{y_1^1, \ldots, y_1^j, \ldots, y_1^{60}, \ldots, y_i^1, \ldots, y_i^j, \ldots, y_i^{60}, \ldots, y_n^1, \ldots, y_n^j, \ldots, y_n^{60}\} \quad (3)$$

where n is the number of frames of the sample data, the i-th frame is represented by the subscript i, and each frame contains 60-dimensional data, the j-th dimension is represented by the superscript j. In motion sample Y, the maximum value a of the j-th dimension max^j is defined as:

$$\max{}^j = y_i^j = \max(y_1^j, \ldots, y_i^j, \ldots, y_n^j) \quad (4)$$

The frame number corresponding to the maximum value is represented by a function:

$$f(\max{}^j) = f(y_i^j) = i \quad (5)$$

In the boundary frame extraction algorithm, the final saved data is not necessarily 120 frames, because the index of the frame may be repeated and the saved data is a set of no duplicate data. Five actions are selected here to display the boundary frame selection results. The results are shown in Table 1.

Table 1. Extraction results of boundary frames.

Action	Number of original frames	Number of boundary frames
Raise right hand and wave	54	35
Wave horizontally	34	26
Raise hands and wave	40	34
Downward punching	31	25
Right hand forward	44	31

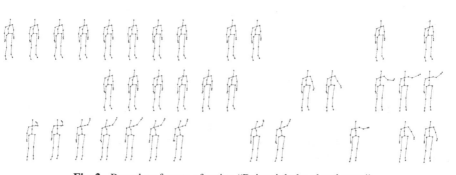

Fig. 2. Boundary frames of action "Raise right hand and wave".

The boundary frames extracted by the experiment "Raise right hand and wave" are shown in Fig. 2. There are three rows of data, 18 frames per row. 35 frames are left after the extraction, which are 1, 2, 3, 4, 5, 6, 7, 8, 10, 11, 16, 18, 23, 24, 25, 26, 27, 28, 31, 32, 34, 35, 36, 38, 39, 40, 41, 42, 44, 47, 48, 51, 53 and 54 frames. Since some non-boundary frames have been deleted, there are some blanks in the figure.

3.2 Extracting Key Frames of Human Motion Data

Based on the extraction of boundary frames, in this section we use frame subtraction algorithm [14] to extract key frames. The frame subtraction algorithm takes the reconstruction error [15] as the subtraction strategy, and subtracts the frame with the least reconstruction error each time.

For the data collected in this paper, we use the Euclidean distance of two frames of data as the measure of reconstruction error. For any two frames y_a and y_b, the distance between them can be expressed as:

$$d(y_a, y_b) = d(y_b, y_a) = \sqrt{(y_a^1 - y_b^1)^2 + \ldots + (y_a^j - y_b^j)^2 + \ldots + (y_a^{60} - y_b^{60})^2} \quad (6)$$

Each frame in formula (6) is regarded as a 60-dimensional vector, and the j-th dimension of y_a is represented by y_a^j. The remaining frames after each subtraction are the key frames currently obtained and the input data for the next subtraction. In the remaining frames, the reconstruction error of the first frame is its distance from the second frame, the reconstruction error of the last frame is its distance from the penultimate frame, and the reconstruction error of any intermediate frame is the larger value of the reconstruction error from itself to the adjacent two frames. Thus, the reconstruction error of any frame can be defined as:

$$e(y_i) = \begin{cases} \max\{d(y_{i-1}, y_i), d(y_{i+1}, y_i)\}, 2 \leq i \leq n-1 \\ d(y_1, y_2), i = 1 \\ d(y_n, y_{n-1}), i = n \end{cases} \quad (7)$$

where n represents the number of remaining frames.

In this paper, we do not need to input the number of key frames when using the subtraction algorithm, but decide whether to continue the subtraction by the symbol of the error derivative, so we need to record the subtraction error each time. The reconstruction error of the k-th subtraction is expressed by $E(k)$, and the second derivative of the k-th can be approximately expressed as:

$$\begin{aligned} dt &= \frac{E(k+1) - E(k)}{(k+1) - k} - \frac{E(k) - E(k-1)}{k - (k-1)} \\ &= E(k+1) + E(k-1) - 2E(k) \end{aligned} \quad (8)$$

where dt is the second derivative of the reconstruction error of the current frame. Since this formula needs to use the reconstruction error of the frame to be subtracted next time, which is not conducive to calculation, we take this one step forward and get the following approximate formula:

$$dt = \frac{E(k) - E(k-1)}{k - (k-1)} - \frac{E(k-1) - E(k-2)}{(k-1) - (k-2)}$$

$$= E(k) + E(k-2) - 2E(k-1) \tag{9}$$

When the *dt* value is negative, we continue to subtract; when the value is positive, the subtraction stops. In order to save memory space, we only reserve the reconstruction error of the first two subtraction frames when computing in the computer. At the same time, we also use a continuous number as a threshold to indicate the number of times that the second derivative is continuously negative. Still using the samples in Table 1, the results after key frame extraction are shown in Table 2.

Table 2. Extraction results of key frames.

Action	Reconstruction error	Sample size	Number of key frames	Continuous number
Raise right hand and wave	7.3611	27	9	1
Wave horizontally	9.1435	27	7	1
Raise hands and wave	4.0653	27	20	1
Downward punching	3.0623	26	11	1
Right hand forward	6.0926	26	11	1

Fig. 3. Key frames of action "Raise right hand and wave".

The result of extracting key frames using the improved frame subtraction algorithm in this paper is shown in Fig. 3. Like Fig. 2, the action is shown "Raise right hand and wave", while continuing the display structure of Fig. 2 (3 rows and 18 columns).

In Fig. 3, the original data is 54 frames, and 35 frames left after the boundary frame is extracted. After the final extraction of the key frames, there are 9 frames leaving, which are 28, 31, 32, 34, 36, 38, 48, 51 and 53 frames. The amount of data changed to one-sixth of the original, but also a good expression of the original action. When using this algorithm, the second derivative calculation of reconstruction error is added, so we do not need to manually input the number of key frames, which avoids the error of key frame selection caused by human factors.

4 Motion Recognition

In this section, the forward-backward algorithm is used to train the key frame-based hidden Markov model (K-HMM). This algorithm is proposed by Baum Welch to train

HMM. It is a special EM algorithm, but it is more called forward-backward algorithm or Baum Welch algorithm because it is designed earlier than EM algorithm. Each training step of the algorithm is divided into two steps: forward and backward. The forward step can be regarded as seeking expectation value, and the backward step can be regarded as seeking the value of the parameters when the expectation is maximized. It is trained iteratively until the parameter converges. First the forward variable is defined:

$$\alpha_t(i) = P(o_1, \ldots, o_t, S_t = i | HMM) \tag{10}$$

Then the backfoward variable is defined:

$$\beta_t(i) = P(o_{t+1}, o_{t+2}, \ldots | q_t = S_i, \lambda) \tag{11}$$

Therefore, the forward probability can be expressed by:

$$\xi_t(i, j) = \frac{\alpha_t(i) A_{i,j} B_{j,o_{t+1}} \beta_{t+1}(j)}{\sum_{i=1}^{M} \sum_{j=1}^{N} \alpha_t(i) A_{i,j} B_{j,o_{t+1}} \beta_{t+1}(j)} \tag{12}$$

The backward probability can be expressed by:

$$\gamma_t(i) = \frac{\alpha_t(i) \beta_t(i)}{\sum_{i=1}^{N} \alpha_t(i) \beta_t(i)} \tag{13}$$

The parameter update formula of the state transition matrix is given by:

$$A_{i,j} = \frac{\sum_{t=1}^{T-1} \xi_t(i, j)}{\sum_{t=1}^{T-1} \gamma_t(i)} \tag{14}$$

The parameter update formula of the confusion matrix is given by:

$$B_{j,k} = \frac{\sum_{t=1, o_k=o_t}^{T} \gamma_t(j)}{\sum_{t=1}^{T} \gamma_t(j)} \tag{15}$$

In order to verify the feasibility of using this method to automatically determine the hidden state number of HMM, the hidden state number of HMM is modified, and more we analyze its recognition accuracy and training time. The results are shown in Fig. 4. Explanation for Fig. 4:

- The input data in the figure is human motion data collected by OptiTrack;
- The abscissa in (a) is the number of hidden states of the HMM, and the ordinate is the recognition accuracy, which is the result of multiplying the percentage by 100 and taking the integer value;
- The abscissa in Figure (b) is the number of hidden states of the HMM. The ordinate is the training time of the model, and its unit is seconds. It can be seen from the figure that the accuracy rate exceeds 90% when the hidden state number of the HMM is 9. As the number of hidden states of the HMM increases, the training time becomes longer and longer. However, when the number of hidden states of HMM exceeds 13,

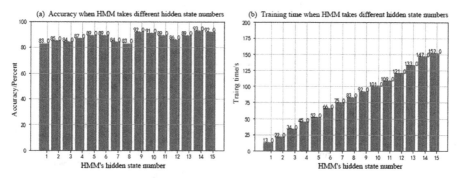

Fig. 4. Accuracy and training time when the model takes different hidden state numbers.

the accuracy rate is higher than that is 9, which may be due to the more detailed classification. Though the accuracy rate is higher, the time cost increases a lot. Therefore, the experiments show that the method in this paper not only integrates training time and recognition efficiency, but also automatically determines the number of hidden states of HMM.

5   System Design

Based on the knowledge mentioned above, in this section, we design a 3D stage interactive environment, in which the interactive function based on motion recognition technology is realized. This interactive system includes five main modules: human motion data real-time acquisition module, virtual environment building module, human motion feature extraction module, human motion recognition module, human-computer interaction module.

For motion data acquisition, OptiTrack motion capture equipment is used to obtain human motion data; for motion feature extraction, we use relative angle as human motion characteristics; for human motion recognition, Gaussian-Mixture Hidden Markov Model is implemented; and for human-computer interaction module, motion recognition and motion segmentation technology are used. The system function module diagram is shown in Fig. 5.

5.1   Real-Time Acquisition of Human Motion Data

In the process of data acquisition, OptiTrack has the problem of data fluctuation and data loss. To solve these two problems, 3 Sigma algorithm of multi-dimensional data and locally weighted linear regression algorithm are used in this paper.

Function and requirement analysis: The key to real-time acquisition of human motion data is real-time data acquisition. The data source is OptiTrack motion capture device; the data structure is selected according to the characteristics of the character model in the 3D environment, and we choose to store the relative offset and relative angle between the user's joints; data storage is to store a complete piece of motion data. In order to enable

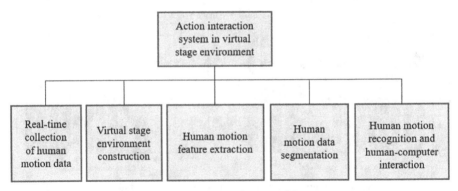

Fig. 5. Function module diagram of the system.

users to intuitively preview their own motion, it is necessary to add a character model in the virtual environment to realize the synchronous movement of the model with the user.

Development environment: The hardware environment requires three steps. Step 1: Set up the infrared camera and fix it. Step 2: Connect the camera to the input of the camera sync device through the cable, connect the output of the camera sync device to the host computer, and connect the camera sync device to the power supply. Step 3: Turn on the host and plug in the software dongle. The software environment includes Motive and Unity 3D. After the Motive software is installed, the T-shaped calibration rod and the right-angle calibration rod are used to calibrate the camera and determine the horizontal plane. Unity 3D is used to complete the development of scenes and functions.

Data acquisition interface design: Fig. 6 is the human motion data acquisition interface. The right side of the figure is the input box of the acquisition, the "collector number" indicates the number of the collector, the "action number" indicates the type of action, and the "num of acquisitions" indicates the times of the action collected. Since we need to wear data collection clothing, and start and end the acquisition are done by clicking the button, it needs someone to assist in the acquisition.

Fig. 6. Human motion data acquisition interface.

5.2 Virtual Environment Construction

In this section, we mainly implement the virtual environment of the stage interactive system, including the creation of models, stage design, camera and light source, audio data and so on. This section is based on the Unity 3D development environment.

Model creation: There are two 3D character models in the scene. One follows the user's movement in real time so that the user can preview his own motion. The other uses the standard data of actors as inputs to restore the actor's movements.

Stage design: Since the stage scene mostly uses red as the main color, we also use red as the main color, and the background uses red and black strip patterns.

Camera and light source placement: Place the camera inside the cylinder and combine the rendering script of the double-sided cylinder to create a circular stage. At the same time, in order to realize the design effect of the real stage, we put a circle of light above the background wall, and choose point light source for the type of light source.

Audio data: In addition to the actions, sound is indispensable in interactive systems. In order to give users a stronger sense of engagement, we also need to prepare appropriate audio data.

5.3 Design of Stage Interactive System

The interactive system mainly includes six parts: 3D character animation reconstruction, real-time human body data acquisition, real-time extraction of human motion data features, real-time segmentation of human motion data, human motion recognition, and motion interaction. After the action recognition is completed, the interactive system receives an instruction. Through different action instructions, the system can complete different interactions. In addition, we also designed some other interactive actions, such as zooming in and out of the character model by outward and contraction of both hands, rotating the character model by drawing circle with the right hand, and shifting the character model to left or right by horizontal sliding of the right hand. The implementation of the system mainly includes the following three steps:

1. Collect actor data as standard input data, then the key frame is extracted by the key frame extraction algorithm, and the reconstruction of the animation is completed by using the interpolation animation reconstruction algorithm;
2. Let the user wear the OptiTrack costume to complete the already trained action in the cave environment;
3. The system will perform feature extraction and motion recognition with the collected user action data, and after the recognition is successful, the corresponding standard action of the actor will be presented, and the corresponding audio will be played at the same time, so that the user can complete the human-computer interaction.

In this way, the system not only allows the user to learn the common actions of the actors, but also helps the user to review the standard actions and correct their actions. The Motive acquisition data interface is shown in Fig. 7, and the final stage interactive system is shown in Fig. 8.

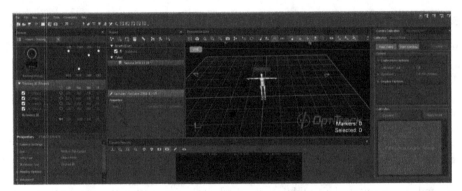

Fig. 7. Motive data acquisition interface.

Fig. 8. Stage action interactive interface.

6   Conclusion

In this paper, we study the feature extraction method of human skeleton data, the human motion recognition method, and the human motion data segmentation method, and finally realize an interactive system under the virtual stage environment. The integrated research can effectively segment human motion and achieve interaction. However, we need to further improve its accuracy in complex scenes. The performance of real-time recognition is also challenged at present. In the future work, we will strengthen the research of image recognition and deep learning, especially for ordinary video data.

Acknowledgments. This paper is supported by Beijing Social Science Foundation (No. 18YTC038), Humanities and Social Sciences Fund of the Ministry of Education (No. 19YJC760150), Beijing Natural Science Foundation (No. 4182018), National Natural Science Foundation (No. 61402016), the open funding project of State Key Laboratory of Virtual Reality Technology and Systems, Beihang University (No. VRLAB2018A05), Beijing Youth Talent Foundation (No. 2016000026833ZK09), NCUT Foundation (No. XN018001).

References

1. Wei, W., An, Y.: Vision-based human motion recognition: a survey. In: International Conference on Intelligent Networks & Intelligent Systems. IEEE (2009)
2. Yang, J., Shi, Z., Wu, Z.: Vision-Based Action Recognition of Construction Workers Using Dense Trajectories. Elsevier Science Publishers B. V. (2016)
3. Liu, T., Song, Y., Gu, Y., et al.: Human action recognition based on depth images from Microsoft Kinect. In: Intelligent Systems. IEEE (2014)
4. Núñez, J.C., Cabido, R., Pantrigo, J.J., et al.: Convolutional neural networks and long short-term memory for skeleton-based human activity and hand gesture recognition. Pattern Recogn. **76**, 80–94 (2018)
5. Min, H., Neve, W.D., Ro, Y.M.: Sparse representation-based human action recognition using an action region-aware dictionary. In: IEEE International Symposium on Multimedia. IEEE (2013)
6. Liang, Y.M., Shih, S.W., Shih, C.C., et al.: Learning atomic human actions using variable-length markov models. IEEE Trans. Syst. Man Cybern. Part B **39**(1), 268–280 (2009)
7. Li, K.: Human action recognition based on fuzzy support vector machines. In: Fifth International Symposium on Computational Intelligence & Design. IEEE Computer Society (2012)
8. Hussein, M.E., Torki, M., Gowayyed, M.A., et al.: Human action recognition using a temporal hierarchy of covariance descriptors on 3D joint locations. In: International Joint Conference on Artificial Intelligence (2013)
9. Baccouche, M., Mamalet, F., Wolf, C., Garcia, C., Baskurt, A.: Sequential deep learning for human action recognition. In: Salah, A.A., Lepri, B. (eds.) HBU 2011. LNCS, vol. 7065, pp. 29–39. Springer, Heidelberg (2011). https://doi.org/10.1007/978-3-642-25446-8_4
10. Subramanian, K., Suresh, S.: Human action recognition using meta-cognitive neuro-fuzzy inference system. Int. J. Neural Syst. **22**(06), 1250028-1–1250028-15 (2012)
11. Wang, J., Qin, X.I., Feng, Q., et al.: Precision testing of OptiTrack system for point measurement. Beijing Surv. Mapp. (2017)
12. Xu, Y., Jun-Yu, L., Feng, J., et al.: A cross-domain collaborative filtering algorithm based on feature construction and locally weighted linear regression. Comput. Intell. Neurosci. **2018**, 1–12 (2018)
13. Xiao, L., Song, M., Zhang, L., et al.: Joint shot boundary detection and key frame extraction. Glycoconjugate J. **16**(11), 2565–2568 (2013)
14. Xiong, W., Lei, X., Li, J., et al.: Moving object detection algorithm based on background subtraction and frame differencing. In: Control Conference (2011)
15. Lin, T., Barron, J.L.: Image Reconstruction Error for Optical Flow (2014)

The Research and Design of 3D Visualization Route Display System Based on osgEarth

Lilong Chen[1(✉)], Guopeng Qiu[1], Jianwen Song[2], and Zhigeng Pan[3]

[1] School of Art and Design, Sanming University, Sanming 365004, China
{20151102,qgp}@fjsmu.edu.cn
[2] China Academy of Art, Hangzhou 310024, China
361688742@qq.com
[3] Digital Media and Interaction (DMI) Research Center, Hangzhou Normal University,
Hangzhou 310012, China
zgpan@cad.zju.edu.cn

Abstract. At present, the display of marching roadmap in the revolutionary history memorial hall are mostly in the form of flat KT board. There are also a few ways of using animation or video, but the way KT board is not conducive to the understanding of the visitors, A certain cognitive burden, animation or video is also not conducive to the latter part of the maintenance, while the lack of interaction. The purpose of this paper is to research and design a marching roadmap system based on osgEarth, which is beneficial to the visitors to understand the revolutionary history more intuitively and to help the administrator to maintain the system content in the future. The system can customize the vector information, provide a highly customized route curve type, route dynamic setting and adjustment.

Keywords: 3D GIS · osgEarth · 3D visualization · Bézier curve · QuadTree

1 Introduction

Fujian Province is a famous old revolutionary area and an important part of the former Central Soviet Area. It nurtures the spirit of the Gutian Congress and the spirit of the great Soviet Area, leaving precious spiritual wealth. Western Fujian is the core area of the former Central Soviet Area and is also one of the starting points for the Red Army's long march. Located in Sanming City in the western part of Fujian, there are a number of national, provincial, and county-level revolutionary cultural relics protection units. In order to further protect red resources, inherit red genes, promote revolutionary traditions, and develop red tourism, red culture continues to flourish, and red genes are passed down from generation to generation. Military museums, revolutionary history museums, red cultural research centers, and patriotism education base and other institutions have been established throughout the country.

Most of the marching road maps used to display the revolutionary history in the museum are still use the regular paper or KT board. Although this method is convenient and cost-effective, it also has many shortcomings: Firstly, it will bring some obstacles to

Z. Pan et al. (Eds.): Transactions on Edutainment XVI, LNCS 11782, pp. 196–205, 2020.
https://doi.org/10.1007/978-3-662-61510-2_19

the cognition of visitors, and it will be difficult to intuitively and clearly understand the context of time and space. Even if there are tour guides, it will not be able to eliminate this cognitive barrier in a short time. Secondly, it is not conducive to the maintenance of the later period. In order to eliminate these deficiencies, the purpose of this paper is to design and research a 3D visualization marching route display system based on the osgEarth framework. The system is intuitive, efficient, and highly customizable.

2 osgEarth Overview and Route Rendering Algorithm Analysis

2.1 osgEarth Introduction

OpenSceneGraph (OSG for short) is a 3D scene graph rendering and scheduling engine based on OpenGL. It is an application programming interface (API) based on the C++ platform. It allows programmers to create an interactive graphics program that is high-performance, cross-platforms more quickly and easily. As a middleware application software provides a variety of advanced rendering features, IO, and spatial structure organization functions.

osgEarth is an OSG extension based on OSG framework that focuses on Digital Earth and has GIS functionality. It has the following features: real-time generation of terrain (texture and elevation), real-time generation based on texture and elevation, support for vector graphics rendering, and flexible vector drawing; support for digital city development; support for terrain scheduling, terrain pre-generation function; provides a wealth of Tools, such as caching, osgEarth Manipulator, etc.; storage space requirements are smaller because there are saved in data servers.

At present, there are a lot of applications based on osgEarth. There are military applications, such as the construction of virtual three-dimensional air battle scenarios [1] and unmanned helicopter mission planning systems [2], etc., as well as applications in engineering, such as three-dimensional water pipe simulation systems. [3], Terrain Modeling [4], Barrier Lake Dam Risk Assessment [5], 3D Urban Scene Construction [6, 7] and Digital Earth, etc. [8], and some navigation applications [9, 10].

2.2 Route Rendering Algorithm Analysis

A smooth curve (the march route map) drawn from a given (selected) set of key coordinate points (latitude and longitude values). This paper's method stitches between Bezier segments. The core of the algorithm must calculate the control points of the Bezier curve. The implementation method is as follows:

There are n + 1 data points $P_i(x_i, y_i)$, $i = 0, 1, 2, \cdots, n$ known on the plane. It is required to use a 3rd degree Bezier curve connection between every two neighboring points P_i and P_{i+1}.

A 3rd degree Bezier curve is determined by 4 points: P_i is its starting point, P_{i+1} is its end point, between the start point and the end point, there are two control points, which are denoted by A_i and B_i in turn. Now we need to determine the two control points.

We take the tangent direction of point P_i as the direction parallel to line segment $P_{i-1}P_{i+1}$ then the coordinates of control point A_i can be expressed as:

$$A_i(x_i + a(x_{i+1} - x_{i-1}), \; y_i + a(y_{i+1} - y_{i-1})) \tag{1}$$

The coordinates of the control point B_i can be expressed as:

$$B_i(x_{i+1} - b(x_{i+2} - x_i), \; y_{i+1} - b(y_{i+2} - y_i)) \tag{2}$$

Where a and b are two positive numbers that can be given arbitrarily. For example, we can take $a = b = 1/4$. At this time, the coordinates of the control point can be obtained by the following formula:

$$A_i\left(x_i + \frac{x_{i+1} - x_{i-1}}{4}, \; y_i + \frac{y_{i+1} - y_{i-1}}{4}\right) \tag{3}$$

$$B_i\left(x_{i+1} - \frac{x_{i+2} - x_i}{4}, \; y_{i+1} - \frac{y_{i+2} - y_i}{4}\right) \tag{4}$$

For example: Let P_{i-1}, P_i, P_{i+1}, P_{i+2} the coordinates of these 4 points be:

$$(x_{i-1}, y_{i-1}) = (1, 1) \tag{5}$$

$$(x_i, y_i) = (2, 2) \tag{6}$$

$$(x_{i+1}, y_{i+1}) = (3, 1) \tag{7}$$

$$(x_{i+2}, y_{i+2}) = (4, 2) \tag{8}$$

According to the formula given above, the coordinates of the control point a can be obtained as:

$$\left(x_i + \frac{x_{i+1} - x_{i-1}}{4}, \; y_i + \frac{y_{i+1} - y_{i-1}}{4}\right) = \left(2 + \frac{3-1}{4}, 2 + \frac{1-1}{4}\right) = (2.5, 2) \tag{9}$$

The parametric equation of the 3rd degree Bezier curves connecting P_i and P_{i+1} is:

$$\begin{cases} x = 2(1-t)^3 + 7.5t(1-t)^2 + 7.5t^2(1-t) + 3t^3 \\ \quad = 2 + 1.5t - 1.5t^2 + t^3 \\ y = 2(1-t)^3 + 6t(1-t)^2 + 3t^2(1-t) + t^3 \\ \quad = 2 - 3t^2 + 2t^3 \end{cases} \tag{10}$$

The first and last segment of this curve cannot be calculated by the above formula, because (x_{-1}, y_{-1}) and (x_{n+1}, y_{n+1}) are used in the formula. These two points do not exist. Use the following processing method:

The value of (x_0, y_0) is used as the value of (x_{-1}, y_{-1}), and the value of (x_n, y_n) is used as the value of (x_{n+1}, y_{n+1}).

That is, in the first segment of the Bezier curve connecting P_0 and P_1, the coordinates of the control point (x_{n+1}, y_{n+1}) are:

$$A_0\left(x_0 + \frac{x_1 - x_0}{4}, y_0 + \frac{y_1 - y_0}{4}\right) \tag{11}$$

In the last segment of the Bezier curve connecting P_{n-1} and P_n, the coordinates of the control point B_{n-1} are:

$$B_{n-1}\left(x_n - \frac{x_n - x_{n-1}}{4}, y_n - \frac{y_n - y_{n-1}}{4}\right) \tag{12}$$

Figure 1 shows a smooth curve drawn using a set of simulation points according to the algorithm (the blue line is the line segment directly connected to each point, compared with the red line).

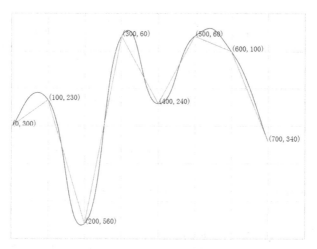

Fig. 1. Smooth curve drawn from a set of simulation points. (Color figure online)

3 System Design

3.1 Development Operating Environment and System Architecture

This system is based on the WIN7 operating system, NVIDIA graphics card, Visual Studio 2010 integrated development environment for the development of single-document applications, the user interface uses Visual Studio 2010 Office style.

The system is based on the rendering engine and integrates third-party library files. The database contains data transmission drivers, data analysis drivers, and data storage drivers. The professional spatial transformation library includes spatial coordinate transformation libraries and spatial data verification libraries. The rendering library includes the underlying OpenGL rendering library, platform-dependent rendering window system libraries, and Platform support library. The system uses the osgdem terrain generation tool also. The system architecture is shown in Fig. 2.

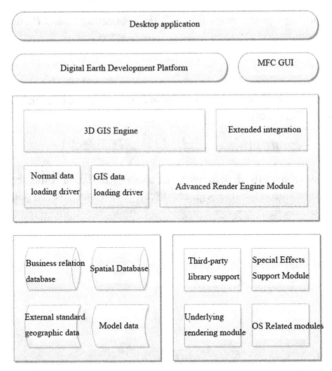

Fig. 2. System architecture.

3.2 High-Resolution Image Acquisition, Processing and High-Precision Terrain Generation

(1) Using 30 m Accuracy Elevation File. This paper uses a 30 m-precision (a pixel represents a 30 m*30 m area) elevation texture file from ASTER GDEM, and specifies the download area by entering latitude and longitude coordinates. Accuracy verification of downloaded elevation files by ERDAS IMAGINE 9.2. Even if the elevation is not very accurate, the elevation can be interpolated later.

(2) Obtaining High-Precision Textures. The resolution of the image texture is more important than the elevation, because the accuracy of the texture directly determines the fineness of the final effect, and the texture cannot be interpolated. This paper uses arceyes Google Satellite Map Downloader to download textures, supports multi-level texture downloads, and automatically populates low-level textures to high-level textures during export stitching.

(3) Terrain Data Preprocessing

Cutting: This system does not require a global elevation, so use gdalwarp (Projection Transformation and Projection Setup Tool) to cut it. The cutting command is as follows:

Gdalwarp -rsc -t srs "+prj=latlong" -te 108 34 109 35 srtm30plus_stripped.tif 10934.tif

Texture coordinate correction and conversion: Firstly, if the texture itself does not have a coordinate system, we need to add a coordinate system to the texture. This paper uses Global Mapper v12.01 to set the earth reference coordinate of the texture image (image correction), and only needs to determine the latitude and longitude data of the four corners of the document. Secondly, if the texture coordinate system is the coordinate system of the UTM plane, it needs to be converted to the geocentric coordinate system. Conversion and interpolation of the coordinate system can be re-projected using ERDAS IMAGINE 9.2. This paper uses cubic convolution to interpolate.

Stitching: This paper uses the ERDAS IMAGINE mosaic tool for stitching.

3.3 Earth File

osgEarth recognizes each function that needs to be completed by reading the earth file. The file is configured by a main tag map, including the high-field elevation, the image texture, and the three commonly used tags that define the cache for the cache and their attributes. The file is defined as follows (Fig. 3):

```
<map name="Globe" type="geocentric" version="2">
    <!--global images-->
    <image name="GlobeImage" driver="gdal">
        <url>../data/image/globe/globel.tif</url>
    </image>
    <!-global elevation->
    <heightfield name="GlobeHeightfield" driver="gdal">
        <url>../data/heightfield/30m.tif</url>
    </heightfield>
    <!--  file cache  -->
    <options>
        <cache type="filesystem">
            <path>./FileCache</path>
        </cache>
    </options>
</map>
```

Fig. 3. Earth file configuration

The digital earth based on osgEarth has the characteristics of dynamic generation, heavy burden, slow scheduling, etc. The solution to slow scheduling is to use caching mechanism to solve problems. On the server side or on the client side, using the osgearth_cache tool to generate the terrain cache in advance, the loading speed at the client startup is significantly increased. The command used is: osgearth_cache --seed china-simple.earth --max-level 3.

3.4 High Definition Image Overlay

There are few images in the Earth's basic framework and the images are quite blurred. This paper uses the following steps to add high-resolution images to the Earth:

(1) Get textures from Google;
(2) Use osgdem for texture correction (re-projection) to generate intermediate files (because Google exports the coordinates of the texture cannot be used directly);

(3) osgEarth loads the intermediate file;

(4) Use the composite tag for texture composition, because the number of top-level image tags is limited, but the composite is not limited, and also to avoid using two image tags directly will result a gap in the joint of two textures.

3.5 Massive Geographical Names Outline Display

This paper uses PagedLOD to dynamically load the landmarks. When the model needs to be rendered in the scene, it will be loaded and read into the program. When there is no need to display, the memory will be removed in time. Read on demand, which saves memory overhead and does not affect system performance during initial load.

Taking Shaanxi as an example, the province has a total of 37,937 place names (including capital city, prefecture-level cities, counties, townships, and villages), which will be saved in the names database (see Table 1). Each name of all provinces, cities and counties, export to an ive file and export it to the corresponding folder, dynamic loading when close to.

Table 1. table_Region, the value of the Lod field in the table is a 4-fold relationship because the relationship between the osgEarth levels at each level is a quad tree-like data structure.

Field	Remarks
PlaceName	
Lod	16, 64, 256, 1024…
Longitude	
Latitude	
IconImage	
Style	

3.6 Shape File Processing

A shape file is a vector file that is used to display roads, border areas, and the appearance of buildings. This paper uses ArcView GIS 3.3 to edit the shape file.

3.7 Increasing the Number of Cities on the Surface

Firstly, the digital city model is built in 3ds max, and the city model is exported through the 3dvri for max plug-in. The exported file format is ive. To add a city to the digital earth, the default is at the origin point, then to move the object to the corresponding point on the surface (in order to cover the surface, we need to set the altitude higher), and then handle the orientation of the object.

3.8 March Route Drawing and Tracking Roaming

The way to set the path is to set the latitude and longitude, height, and finally convert the absolute coordinate point into a world coordinate point.

The information of the key points in the route is composed of coordinates (longitude x, latitude y, height z) and speed (w). The route is obtained according to the key point information, and the tracking is finally set. Table 2 is the march route database table.

Table 2. table_Route.

Field
Name
Description
Speed
Width
Color
Opacity

The key point database table design is shown in Table 3.

Table 3. table_Location.

Field
Position
Name

In the path roaming, the orientation of the object at the key point n is calculated based on the connection between the n point and the n-1 point.

According to Eqs. (3) and (4), the pseudo code for drawing routes in canvas is as follows:

```
begin
    Input coordinate point array points;
    Input line style, line color, line width;
    For i from 0 to point.length;
    repeat
      If i==0 moveTo the 0th coordinate point;
      Else, input points, i-1 to get control point P;
            input P.pA.x, P.pA.y, P.pB.x, P.pB.y,
            point[i].x, point[i].y;
end.
```

4   Tests and Results

It takes a few minutes to run for the first time, as the terrain needs to be generated dynamically based on the image and elevation files, and the cache files will be generated in the specified path. After that, opening will be very fast (Fig. 4).

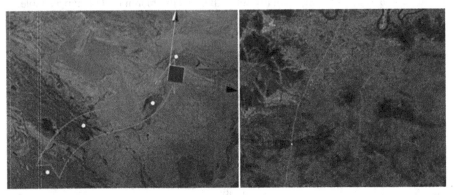

Fig. 4. Left figure (Route Control Point) is a preview of dynamically generated Bezier route while adding control points. After adding a series of control points, the simulated marching route (visualized tour route) was drawn according to the algorithm, and was shown in right figure (Roaming Path). The smooth Bezier curve achieved our intended purpose.

5   Discussion

This paper studies and designs an osgEarth-based 3D visual marching route display system, which can customize vector information, provide highly customized route curve types, and dynamically set and adjust route, drawing a smooth Bessie according to a given series of control points. The curve is used to simulate the march route. The system is used in the revolutionary historical memorial. It can help the visitors to understand the revolutionary history more intuitively, and it also helps the administrator to maintain the content of the system later. The system is still in the preliminary stage of design and development. In addition to the global bird's-eye view mode, it plans to add multi-perspectives and auto-roaming modes in accordance with the selected route. There are still a lot of works need to be done in the physical hall.

Acknowledgments. We would like to thank the support of the Fujian Young and Middle-aged Teacher Education Research Project with project codes JAT160472. Also, we would like to thank the Fujian Science and Technology Agency for funding support of this research through Soft Science Project with project codes 2017R01010181.

References

1. Wang, L., et al.: Constructing of large scale 3D air-battle scene based on osgEarth. Comput. Eng. Softw. **37**(01) (2016)
2. Huang, F.G., et al: Missions plan system development of unmanned helicopter based on osgEarth. Helicopter Tech. (187) (2016)
3. Qiu, W.S., et al.: Design and implementation of 3D water pipe network system based on osgEarth. J. Henan Sci. Technol. (10) (2014)
4. Li, W., et al.: Terrain modeling simulation based on osgEarth. Data Base Tech. (01) (2014)
5. Zhu, J., Wang, J.H.: Interactive virtual globe service system based on osgEarth. Appl. Mech. Mater. **340**, 680–684 (2013)
6. Xiaodong, W.U.: Construction of urban 3D scene based on osgEarth. Geospatial Inf. (2013)
7. Yan-Chao, Y.U., Han-Wei, X.U., Xiao-Dong, W.U.: Research of the organization and scheduling of feature model of urban three-dimensional based osgEarth. Geomat. Spatial Inf. Technol. (2014)
8. Chen, B., Ren, Q., Yang, H.: Design and realization of 3D platform for digital earth based on osgEarth. Electron. Sci. Technol. (2015)
9. Wu, X.X.: Study on Virtual Campus Roaming and Information Display System Based on osgEarth. Dalian Maritime University (2013)
10. Wang, J.H., et al.: Research on virtual campus scene modeling based on osgEarth. Geomat. World (01) (2011)

Forensic Analysis of Play Station 4

Feng Gao[1], Ying Zhang[1], and Hong Guo[1,2(✉)]

[1] The Third Research Institute of Ministry of Public Security, Shanghai 201204, China
{gaofeng,zhangying}@stars.org.cn
[2] Academy of Forensic Science, Shanghai 200063, China
guoh@ssfjd.cn

Abstract. As we known, video game consoles are not only the carriers of games to global users, but also the communication and social platforms due to their embedded instant messenger software. In addition, the data structure and storage of game consoles enables them to provide facilitation to criminals who are willing to conceal potential evidence. As the eighth-generation of video game consoles, demand for play station 4 was strong, which means that it attracts numerous users. Nevertheless, seldom researchers conducted analysis towards this console. This paper provides a method for acquiring digital evidence on play station 4 and gives a detailed description of embedded instant messenger software.

Keywords: Play station 4 · Digital forensic · Video game console · File system

1 Introduction

The play station 4 was developed in 2013 by Sony Interactive Entertainment. Compared to the Cell microarchitecture utilized in play station 3, the new console adopts an AMD-based X86-64 architecture, which facilitate the development of games. Except for discrepancy in architecture, play station 4 places an increased emphasis on social interaction, including playing games online with friends or controlling games remotely [1]. Besides that, game players could upload gameplay video to social media, or communicate with players from all over the world. Another point it was acclaimed is that play station 4 was not imposing the restrictive digital rights management schemes like other consoles such as Xbox one [1]. Attributed to those improvements mentioned above, there were more than one million pre-orders in August 2013 [2]. Until the end of 2018, Sony confirmed that it had sold-through 91.6 million play station 4 units.

Despite numerous customs of the console, little research or information about investigations in forensic manner was found. This paper aims to give a brief description of hardware as well as provide a reasonable scheme for consoles from the aspect of forensic investigation.

For the scope of this paper we will give a detailed introduction of play station 4 from hardware and software both, including demonstration of hard drive, configuration of console and playstation network, analysis of operating system and file, as well as

Z. Pan et al. (Eds.): Transactions on Edutainment XVI, LNCS 11782, pp. 206–214, 2020.
https://doi.org/10.1007/978-3-662-61510-2_20

on-site investigation. Section 2 will describe previous research concerning related game consoles. Section 3 will describe partition layout, file system, system data and so on. More analysis of system files will be given in this chapter as well. Conclusion is in Sect. 4.

2 Literature Review

Previous investigators have conducted numerous research concerning various video game consoles, including Xbox 360, Xbox one, Play Station 3 and so on. Due to their unique operating system, file system and data, researchers were confronted with severe technical challenges. Research about typical consoles would be demonstrated as follow.

2.1 Xbox 360

Previous work conducted by Steven Bolt [3] has shown a detailed interpretation on Xbox 360, including hardware and software both. Concretely speaking, the research introduced the design of console, system hardware structure as well as system disposal. Xbox Live and system configuration were described as well. As for forensic examination, it demonstrated file locating on the drive, and then a further research about Xbox 360-specific file types was conducted. In addition, analysis of gaming artifacts, cache folder and content folder was given in details, as well as network traffic analysis.

2.2 Xbox One

Jason Moore and other researchers [4] have committed the study about Xbox one in the past few years. They proposed a new method of examining an analyzing Xbox one, imaging hard drive, viewing partition lay out, computing and comparing hash value of files under root directory as well as analyzing specific files locating on the drive. It also showed some Xbox one-specific file format such as. xvd and gave an elementary interpretation regarding those files. Network data captured during the whole experiment was analyzed and relevant clues were found as well.

2.3 Play Station 3

Research conducted by Scott [5] concentrated on performing various operations to hard drive in the play station 3 and examining corresponding result, which are constituted of control boot test, write blocker test, OS installation test, imaging over a network test, game OS reinstallation test, backup utility test, hard drive swap test, browser test and hard drive decryption test. Through above tests, it is found that operating system in play station 3 is unable to be read, and traditional methods such as data carving could recover data from this console only.

2.4   Play Station 4

Previous research [6] proposed a method for acquiring data from play station 4 in a forensic manner. Traditional forensic measures including data carving, web browser, bookmarks, history and recent items examination, time and date test, drive backup and restoration, offline and online write blocker test, shadow drive online web browser analysis, playstation network analysis, USB upload and download test and so on. After those attempts mentioned above, the research proposed the best practice methodology for the forensic analysis of a play station 4.

3   Initial Analysis of File System

In order to commit an analysis in detail, hard drive is supposed to be removed from the console, while certain measures are required to be taken for the sake of avoiding damage to hardware. Then we are going to image the whole drive, which enables researchers to repeat experiments without worry about data irreversibility due to errors and failures. Finally an initial analysis of file system would be demonstrated as follow.

3.1   Hard Drive and Image Acquisition

As it is known, relevant data for play station 4 is stored in a hard disk locating in the machine. After opening the top cover and disassembling the corresponding screw, the hard drive could be removed. It is noticed that there may be some differences between old and new models while disassembling the console.

Then the hard drive will be imaged via using professional tools. In our experiment, Forensic Falcon will be utilized to make the image. It is an imaging solution associated with write-blocker [7]. Firstly, we select the 'hard disk to file' mode on the equipment and then connect the disk removed from game console to the read-only interface to ensure that it could be recognized. In the next step, the extension of image file would be selected as well as other configurations. Besides that, the disk is required to be hashed to preserve the integrity of data. Figure 1 shows the process of committing the image file for game console.

3.2   Initial Analysis

In this chapter, the tool called X-ways Forensics would be utilized. It is a forensic tool which is capable of reading partitioning and file system structures inside raw image files, automatic identifications of lost or deleted partitions and so on [8].

After above operations, the image for this game console is acquired with extension .001. It is found that the hard drive is composed of 15 partitions. Details of partitions are listed in Table 1. As shown the file system in play station 4 is not standardized system such as NTFS.

We use X-ways Forensics to examine content of each partition and find it could not be interpreted even with professional tools. Then key words including user name, password, network name and game name were utilized for simultaneous search over the hard disk. Unfortunately, nothing was found after that operation.

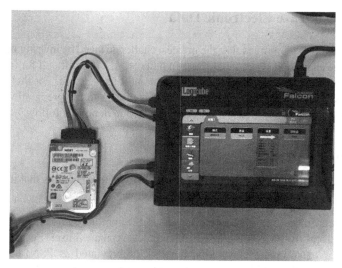

Fig. 1. Image the hard drive via using Falcon

Table 1. Partition for Play station 4.

Name	File system	Total capacity
Partition 1	unknown	1.0 GB
Partition 2	unknown	16.0 GB
Partition 3	unknown	420 GB
Partition 4	unknown	8.0 GB
Partition 5	unknown	1.0 GB
Partition 6	unknown	1.0 GB
Partition 7	unknown	1.0 GB
Partition 8	unknown	1.0 GB
Partition 9	unknown	8.0 GB
Partition 10	unknown	512 MB
Partition 11	unknown	1.0 GB
Partition 12	unknown	16.0 MB
Partition 13	unknown	128 MB
Partition 14	unknown	1.0 GB
Partition 15	unknown	6.0 GB

4 Capture of Live Electronic Data

Different from other game consoles, the greatest challenge faced by investigators for play station 4 is the encrypted system it utilizes. In other words, it is difficult for investigators to commit in-depth analysis of operating system and file structure. For this reason, the most effective way is to conduct an on-site acquisition. And the preparation work and concrete procedures committed will be demonstrated in details as follow.

4.1 Preparation

Before the acquisition starting, sample data should be made. Features that investigators will be interested in are described and made as follow.

User Account. In order to obtain relevant data, different accounts are supposed to be created. The local user account consists of three users, named user1, user2 and user3. In addition, we also create playstation network (abbreviated as PSN) accounts and subscription-based PS PLUS account so that all features would be included.

Messages. In order to test messages in the console, it is necessary to add contacts to the users' buddy list first. Besides that, personal or group messages are likely to be sent to users created as well.

Error History. It is found that in the experiment some network-depended configuration would lead to errors and those errors would be added to the log file. Therefore, we closed the network connection, recorded the access time of record and then checked if there was any error generated.

Internet Browser. According to previous research [9], the internet browser can store 100 websites and 8 most recently visited webpages. To verify this point, 103 websites and 103 bookmarks were prepared, including pictures, web pages, websites and duplicate entries.

4.2 Live Acquisition

After preparation, live acquisition of digital evidence would be conducted by researchers. The point is that digital evidence is required to be captured and preserved carefully so as to ensure its authenticity and integrity. Therefore, various features are supposed to be considered as demonstrated as follow.

Date and Time. Timestamp is an important factor affecting the acquisition so that most investigators show solicitude for related data. Such information is likely to be related to the specific time of crime. Nevertheless, timestamp may be modified deliberately by the criminals. In this situation, investigators are required to obtain such information as much as possible. According to previous research [6], majority of features such as Trophies provided time information while others including web browser do not.

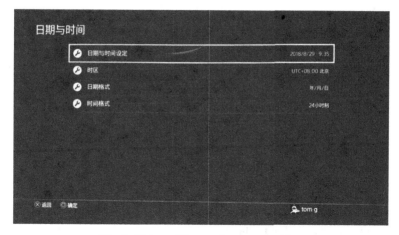

Fig. 2. Date and time under system setting

The timestamp of current system would be demonstrated in 'Date and Time Settings' as shown in Fig. 2. It is noted that off-line access is utilized usually to ensure that data would not be modified.

Besides that, we can also find timestamp-related information elsewhere. Messages and error history records time information of each record in details as shown in Fig. 3. System date and time could be acquired effectively with the help of such information listed above. As a result, when committing a digital forensic investigation, not only the timestamp under system settings but also information located in different features is required to be captured.

Fig. 3. Timestamp information under messages and error history

System Information. As we known, game consoles are no longer game platform but computer-like machines. Under this situation, information concerning system is supposed to be acquired. Concrete information is listed in Table 2.

Table 2. System information.

Number	Name
1	Console name
2	Operating system version
3	IP address
4	MAC address (LAN)
5	MAC address (Wi-Fi)
6	Other related system information

Those information is central to investigation of play station 4 and could be obtained under 'System Information' in the laptop.

Web Browser. Web browser is pre-installed application of play station 4 so that customers are able to access to major websites. Analysis of browser history is a common and critical part of forensic investigation. It aims to examine and analyze the information what suspects viewed on the internet through a web browser. As we known, some critical evidence could be hidden in online search and web browsing history. Web history is helpful for finding or excluding suspects in cases involving pornography or fraud. Even in other cases, web history is able to determine the outline of criminals' behavior. It is know that suspects who use play station 4 usually leave clues about their browsing websites and search keywords in web browsers. Therefore it is of great significance to commit the research on extraction and analysis of web history.

Frequent access to certain websites is an important part of suspects' personal behavior. Most frequently-visited 10 websites by users contribute nearly 80% page views. Thus most investigators treat frequently-visited websites as an important indicator while describing suspects' behavior. In addition, search keywords would be taken into account as well. Analysis of web history can help investigators understanding above information quickly. Via selecting 'browsing history' all web histories will be demonstrated as shown in Fig. 4.

Fig. 4. Web browser history

Bookmarks. Similar to web browser history, bookmarks reflect suspects' personality characteristics. Bookmarks examination aims to analyze the information they are interested in and helps finding potential evidences from websites. In addition, timeline of bookmarks is also critical for investigators for the reason that it may be related to period of crime.

Bookmarks could be acquired via selecting 'bookmarks' after 'internet browser'. Then all information of webpages which were bookmarked by specific suspects and were sorted by timeline.

Recently-Visited Websites. Recently-visited websites describe the last webpages accessed by the suspect. It is known that most investigations are required to be conducted under an offline environment. In such situation, 8 websites would be demonstrated via selecting corresponding pages.

Error History. When play station 4 encounters various bugs, the console will record error codes, error numbers and occurring date in the error history. And up to 100 records can be displayed.

Such information could be obtained via entering 'system' firstly and then selecting 'error history'. Then error code, error information in details and corresponding date and time will be demonstrated.

Buddies. Play station 4 allows users to add common friends as well as add friends from facebook. Thus investigators can acquire relevant communication information from buddy list. Since the real name is required in communication, investigators could find all users' real name in the conversation. According to previous research, up to 2000 buddies can be added for one user.

After sending request to players, users will be accepted as buddy. Then each buddy's information could be viewed, including his PSN login status and the game he is playing. Two players becoming buddy will follow each other automatically. Moreover, buddy's activities and broadcasts will be displayed on the latest news.

Players could see the latest activities and trophies on the interface of buddy's profile. Besides that, users can send messages or party invitations through this interface. The point is that buddy list could not be obtained in offline mode. While accessing to internet, buddy list could be found in 'follow' under 'buddy'.

Messages. Play station 4 provides communication service to users. And user group could be established between players and up to 8 users could compose a group. Players in one group could communicate with each other. Some criminals are likely to utilize this function to communicate criminal activities. Thus message extraction is a critical part of investigation.

Different from Buddy list, conversation between players in one group could be displayed even in the offline mode. That means messages were cached in the console when they were received. Via selecting 'Message' we can find all conversations between buddies.

User Profile. Unique number and personal data would be demonstrated in user profile, which helps investigators digging out the characteristics of suspects. Concretely speaking, real name, personal avatar, virtual image and other personal information would be listed in user profile. And certain privacy settings such as what kind of people are allowed to view user's activities and trophies are stored in this interface as well.

User profile must be obtained in the online mode. Then the player's status, trophies, personal data and other information could be viewed clearly.

5   Conclusions

With the development of internet technology, various game platforms have emerged and game users increased as a result. The criminals including extortion and identity theft via using video game consoles have grown with the upgrade of function and performance of consoles. As the eighth generation of game console, play station 4 should not be underestimated. Nevertheless, few research have been conducted on play station 4, which resulted in the fact that little was known about operating system, system composition, disk partition and other information. Besides that, there are still lack of relevant basis and unified standards in the field of video game consoles forensics. Thus the formal examine and analysis method of play station 4 is supposed to be proposed as soon as possible.

In this paper we demonstrate a concrete analysis of play station 4, which includes preliminary analysis of partition, system file examination, key words research as well as live acquisition. Through above analysis, it is found that file system is encrypted, so system files could not be interpreted and key words could not be found as well. Therefore for play station 4, we propose a new method for live acquisition, which requires to obtain information about date and time, system information, web browser, bookmarks, recently-visited websites, error history, buddies, use profile as well as messages. More exploration of play station 4 would be conducted in the future work.

Acknowledgments. This paper is supported by National key research and development plan, The People's Republic of China ministry of science and technology, project number: 2017YFC0803805.

References

1. Information on. https://en.wikipedia.org/wiki/PlayStation_4
2. Information on. https://blogs.wsj.com/digits/2013/08/20/sony-says-playstation-4-launches-nov-15/
3. Steven Bolt: XBOX 360 Forensics: A Digital Forensics Guide to Examining Artifacts (2011). ISBN 978-1-59749-623-0
4. Moore, J., Baggili, I., Marrington, A., Rodrigues, A.: Preliminary forensics analysis of the Xbox One. Digit. Investig. **11**, S57–S65 (2014)
5. Conrad, S., Dorn, G., Craiger, P.: Forensic analysis of a Playstation 3 console. In: Chow, K.-P., Shenoi, S. (eds.) DigitalForensics 2010. IAICT, vol. 337, pp. 65–76. Springer, Heidelberg (2010). https://doi.org/10.1007/978-3-642-15506-2_5
6. Davies, M., Read, H., Xynos, K., Sutherland, I.: Forensic analysis of a Sony Playstation 4: a first look. Digit. Investig. **12**, S81–S89 (2015)
7. Information on. https://www.logicube.com/shop/falcon/?v=1c2903397d88
8. Information on. http://www.x-ways.net/forensics/index-m.html
9. Information on. http://manuals.playstation.net/document/gb/ps4/browser/bookmark.html

Interactive Experience and Communication of Public Service Advertisements Based on User's Subconscious Behaviors

Jia Wang and Hong Yan[✉]

HaiNan University, Haikou 570228, China
1907556443@qq.com, yanhong@hainu.edu.cn

Abstract. Traditional public service advertisements mostly adopt one-way pub-licity, such as radio and television, which has limitations on them. With the rapid development of communication technology, public service advertising has a new mode of publicity, such as interactive advertising, incentive advertising. Based on the development status and problems of public service advertisements, this study takes care of the mental health of depressed patients as the starting point and expands the communication experience of public service advertisements from the perspectives of psychology and behavior. On the basis of analyzing the "iner-tia" of users' subconscious behavior, a button model is designed to guide users' subconscious behavior. By simulating the negative effects of users' subconscious behaviors, such as bringing pain to depressed patients or the possibility of suicide, this study gives people a self-reflection and self-evaluation experience in a natural interactive way, so as to achieve the communication significance of public service advertisements.

Keywords: Subconscious behavior · Depressive disorder · Public interactive advertising

1 Introduction

As an important way to spread social public awareness, public service advertisements often lead more people to pay attention by self-reflection and self-criticism, so as to achieve in-depth communication effect. For example, in 2017, 170,000 h are used on China's radio and television public service advertisements production and 630,000 h on broadcasting [1]. However, even with such amount of production and broadcasting time, it still only has certain transmission effect.

For example, despite there are many public service announcements about depression on television, public awareness of depression remains limited [2]. The research [3] on a questionnaire survey of 207 random people found that 57.5% of respondents believed that depression was a sign of personal weakness. 65.7% of respondents considered it was best to avoid contact with people with depression. 41.1% of respondents believed depression was not a real illness. There is a huge gap in the public's understanding of depression patients and a stigma towards depression, which is partly responsible for the

© Springer-Verlag GmbH Germany, part of Springer Nature 2020
Z. Pan et al. (Eds.): Transactions on Edutainment XVI, LNCS 11782, pp. 215–222, 2020.
https://doi.org/10.1007/978-3-662-61510-2_21

Fig. 1. Hair blown by the wind. (source: https://www.madmoizelle.com/chimiotherapie-cancer-publicite-interactive-291938)

increasing number of people suffering from depression who commit suicide or do not seek medical treatment every year.

On the other hand, in 1908, Simmel mentioned the "social interaction theory" in his book "sociology", which mainly studied how people would react under specific circumstances, aiming to solve the profound problems of society by scientific and effective analysis of social problems in development and targeted design [4]. With the rapid development of science and technology and the popularization of intelligent products, interactive public service advertisements begin to appear in the public's vision. For example, the public service advertisement "hair blown by the wind" in the subway station of children's cancer foundation in Sweden as shown in Fig. 1 is a good example. The Swedish children's cancer foundation utilized sensors to sense the train's location. When the train passed by, the model's hair was blown dynamically. While everyone was enjoying the advertisement happily, the model's hair was blown off, and everyone's expression changed from surprise to meditation. The slogan of this public service advertisement is "Every day a child is diagnosed with cancer". This example shows that interactive advertising makes the content more interesting and emotional.

The purpose of this study is to design the viewing experience of public service advertisements for caring depression patients' mental health. By the interactive way to improve the attraction and appeal of public service advertising, achieving the more popular dissemination. The design of interactive public service advertisements mainly focuses on the behavior of "people" themselves. By observing people, design how to interact [5]. Freud's hierarchy of consciousness theory states that a person's mental activity consists of three distinct levels of consciousness: consciousness, subconscious and unconscious. In most interaction design studies [6], the user's conscious behavior is more concerned. The user's conscious behavior is the subjective experience that comes to mind at will and is clearly perceived [7]. Therefore, these designs are often aimed at the user's subjective needs generated by the behavior. Public service advertisements can hardly become users' subjective needs, so interactive public service advertisements have great limitations in utilizing users' conscious behaviors [8].

Compared with conscious behavior [9], unconscious behavior is characterized by objectivity, instinct, reflexivity, empiricism, imperceptibility and non-target guidance. In the contact process between the user and the product, different from the purposeful operation behavior [10], the subconscious behavior is the behavior based on the past experience and habits, which does not require brain thinking but "inertia" to make choices or judgments and so on. For example, when people are nervous, they scratch their head; when they sit for a long time, they shake their legs.

This study aims to use the "inertia" characteristics of users' subconscious behaviors to design and guide users' subconscious behaviors, and to simulate the possible negative effects, so as to give people a self-reflection and self-criticism experience in a natural interactive way and achieve the communication effect of public service advertisements.

2   Related Works

Interactive advertising is a way of communicating between people and the media. The interaction establishes a bridge for the viewers to participate in artistic creation, so that the viewer can really get close to the connection between media art and design works, and at the same time use the emotional satisfaction of being valued and self-selected to respect human rights. With the development of network technology, modern technology and information have entered a new stage. The form of digital media is very broad and its expressions are becoming more diverse. The dynamic posters in the digital media era are supported by digital technology, digitizing the original information and expressions, making them accurate and efficient.

In 2016, MaseReor, a six non-profit charity in Germany, designed a set of dynamic public welfare posters called "Social Swipe Cards". Since 1959, it has been active on the international stage to address the hunger and imprisonment of children in impoverished areas of the world. There are bundled hands or delicious bread in the interactive public service advertisement. Every time you swipe your card, you not only cut the bread on the poster or the rope that binds your hands, but also donated 2 euros to the poor. When the credit card data is verified, the bread in the billboard will be taken away, and the words of thanks will appear. This sense of sight makes people feel that the donation seems to be effective immediately (Fig. 2).

In 2014, Swedish hair care brand apolosophy erected a billboard in the subway station in the capital city of Stockholm was mentioned above. The billboard was equipped with a Raspberry Pi micro-computer. The picture shows a young girl with hair full of hair, using the sensor device, when the tram After the passage, the wind will be blown up by the wind, which is very suitable for the product. In fact, this is a public service advertisement for caring for cancer patients. When the hair goes with the wind, the girl reveals the head of chemotherapy. It can be seen that the application of new media in foreign countries is very mature, and the creativity is also endless.

Fig. 2. Dynamic public welfare poster designed by charity MaseReor for "social swipe" (source: http://www.sohu.com/a/115242700_501142)

3   Use of Subconscious Behaviors

Subconscious behavior is the instinctive behavior of human beings. It is above intuition and before consciousness. These instinctive behaviors are derived from human vision, hearing, taste and other feelings based on human senses. For example, if a person places a red button in a particular environment, most people will choose to press it. In the same way, when the fire alarm sounds, people will run to the survival channel. When encountering danger, this instinct to survive will make their body urgently act. We observed in the observation that the user population will make subconscious interactions with the red button. Figure 3 is the appearance of the red button.

Fig. 3. The button that causes subconscious behavior (Color figure online)

The purpose of advertising is to attract people's attention and interest in content more easily and easily. Traditional public service advertisements are displayed in a specific space and time, and lack of interaction with users. Different from traditional public service advertisements, the audience can get more information. It is even possible to build a virtual world that allows viewers to come into contact with products or information in the virtual world, leaving a deep impression and reaching depth.

4 Design and Implementation

The code for the interaction of the work is to write the interactive language through Processing, and the externally connected 3D somatosensory camera Kinect. This advertisement has a total of four videos, one for the first, two for the self-reported video of the depressed person, and one for the end of the subconscious behavior. The venue for public interactive advertising is suitable for areas with rest and waiting for people. Such as subway stations, bus stations, street crossings and other areas where other people are more mobile (as shown in Fig. 4).

Figure 5 shows a partial photo of the interactive experience of PSAs for practical application. As shown in Fig. 5A, when the user enters the recognition area of Kinect, the video starts playing. As shown in Fig. 5B, when the user leaves and watches again, the video advertisement content is switched, and the self-report of the male depression patient is randomly converted into the female depression patient, which successfully attracts the attention of the viewer. As shown in Fig. 5C, the user noticed the red button, the subconscious press, the self-reported video of the depression patient stopped, the screen turned black, the music rang again, and the user's self-reflection prompted the video of the depressive patient.

Fig. 4. Applications

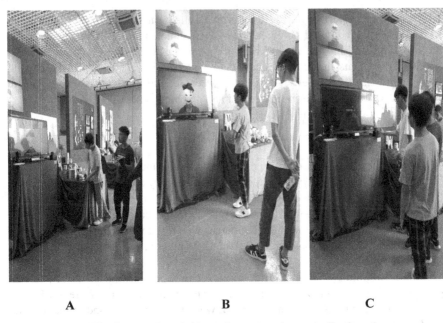

A   B   C

Fig. 5. Part of user's interactive process in our application.

By observing the subconscious behavior in life, combining the themes of the thesis, mining the psychological effects, instinctive reactions or emotional hints behind the user's behavior, the works transform these data into meaningful interactive resources. Through the subconscious behavior derived from life, processing and participation in the public service advertising interaction, the interaction itself is convenient and more affinity and attractive, thus achieving the purpose of interactive experience and communication.

Practice has proved that one-on-one inquiry is the best way to collect psychological activities behind subconscious behavior. Through skillful observation methods, the study of human behavior, the analysis of motivation behind behavior, and the general classification and summarization of results are important ways to transform into effective interactive resources.

5 Contents

Copywriter: Self-report of patients with depression the self-report is described in the first person's tone, giving the viewer a close feeling experience in the advertisement, which can bring the distance to the viewer. Through a series of self-reports, it can clearly describe the inner activities and feelings of patients with depression, so that the viewer can more intuitively understand the life status of patients with depression, more realistic, and more likely to resonate with the viewer.

Props: Smile Mask was shown in Fig. 6. There are three main reasons for designing masked shooting: (1) The film is targeted at people with depression, because most people have a strong sense of shame for depression, so mask objects are used for the protection of the subject. Smile depression is also a kind of depression. (2) Although some patients feel extreme pain, depression, sorrow and sorrow in their hearts, their appearance is as if they are nothing, with a smile. (3) It has a certain attraction in the expressiveness of advertising, which leads to curiosity and can also bring about an active atmosphere to a certain extent.

Fig. 6. The Smile Mask for depression patient protection design.

Sound Processing: There are two main reasons for designing sound processing to make sound changes. (1) The film is targeted at people with depression, because most people have a strong sense of shame for depression, and the sound is processed for the protection of the subject. (2) Public service advertisements for depression are generally very depressed and are not acceptable to most people. The processed sound has certain appeal and interest in advertising expressiveness, which leads to curiosity, more in line with the trend of contemporary society, and can also bring the atmosphere to a certain extent. The background music in the head section of the advertisement uses the humming sound of the TV interference. This music gives people a feeling of irritability and depression, which echoes the living conditions of the depressed patients. The use of background music in the video content section is a feeling of sorrow and sorrow, and there is a sense of substitution for the viewer. After the button is taken, the background music of the video part uses the fast and slow rhythm, which makes the picture match the music, giving the soul a shock and reflection.

Video Editing: The title video is created by the AE and Pr. The snow-screen effect made by AE is then dubbed in Pr. The advertising content is divided into two parts. The propaganda part is a self-report for two depressed patients, and the random form is played during the interaction. The video of the advertisement content is divided into two partial clips, one is the abnormal state and the normal state of the depressed patient. Part of the reflective behavior is the use of the reporting form of the current state of depression and the switching lens of the normal life of the depressed patient. The background effect of the video is the black halo background effect, which is intended to make the self-reporting monologue effect of depressed patients more realistic (Fig. 6).

6 Conclusions

Researchers [11–15] have a high degree of consistency in the point that "subconsciousness is related to people's prior experience." People's understanding of some things and the acquisition of knowledge are derived from the early memories in a subconscious process. In the process of cognition, people are stimulated by some information from prior experience or memory. This information is not necessarily the same as the current situation, but only relevant. By using people's subconscious behavior, and then expressing the harmonious relationship between people and things, it is an effective way to spread public service advertisements.

When designing a brand-new product, this memory of object sharing is introduced into it. Since any event may contain something that we all know or share, using the core elements of this deep design, we can trigger similar behaviors when the external representations are different. By analyzing the "inertia" characteristics of the user's subconscious behavior, designing the subconscious behavior that guides the user, and simulating may have a negative impact. In the natural way of interaction, it gives people a self-reflection and self-evaluation experience, so as to achieve the spread meaning of public service advertisements.

References

1. China Central Television. http://www.cctv.com
2. Manfred, L.: Usability engineering. Inf. Technol. **44**(1), 3–4 (2002)
3. Wen, L., Pan, S., Tang, S.: Analysis of public attitude towards depression and its influencing factors. Chin. J. Health Psychol. **25**(05), 670–673 (2017)
4. Salomon, G.: Distributed cognitions: psychological and educational considerations (1993)
5. Sohn, M., Nam, T., Lee, W.: The impact of unconscious human behavioron sustainable interaction design (2009)
6. Preece, J., Sharp, H., Rogers, Y.: Interaction Design: Beyond Human Computer Interaction (2015)
7. Streitz, N., Magerkurth, C., Prante, T., Röcker, C.: From information design to experience design: smart artefacts and the disappearing computer. Interactions **12**, 21–25 (2005)
8. Liu, P., Su, K.: A research on the emotional cognitive model based on user subconscious behavior. In: 2014 IEEE Workshop on Electronics, Computer and Applications (IWECA). IEEE (2014)
9. Lv, W.: Interactive design research based on action-oriented and subconscious. Art Des. **12**, 96–98 (2015)
10. Suri, J.F.: The experience evolution: developments in design practice. Des. J. **6**, 39–48 (2001)
11. Nelissen, J.M.C.: Intuition and problem solving. J. R. Coll. Gen. Pract. **28**(28), 71–74 (2013)
12. Blassnigg, M.: Documentary film at the junction between art, politics, and new technologies. Leonardo **38**, 443–444 (2005)
13. Li, L.: The study on human-computer interaction design based on the users' subconscious behavior. IOP Conf. Ser.: Mater. Sci. Eng. **242**, 012125 (2017)
14. Vetulani, J.: Biology of memory. Neurologia i Neurochirurgia Polska **33**(Suppl.), 1–19 (1999)
15. Li, J., Salcuni, S., Delvecchio, E.: Meaning in life, self-control and psychological distress among adolescents: a cross-national study. Psychiatry Res. **272**, 22–129 (2018)

National Culture - The Soul of Chinese Animation

Xiaoli Dong[⊠]

Jilin Animation Institute, No. 168 BoShi Road, Changchun, People's Republic of China
417463493@qq.com

Abstract. National culture is the soul of Chinese animation. Chinese culture has a long history, which accumulates a wealth of material for the creation of animation scripts, and also provides inexhaustible role images for the shaping of Chinese animated characters. To revitalize the Chinese animation industry, it is necessary to inject the soul of Chinese national culture into Chinese animation, which requires China's animators to improve their ideological understanding, strengthen the study of national culture, reform the current animation education system properly, have the courage to practice and explore, and then make Chinese animation a distinctive seal.

Keywords: Chinese culture · Chinese animation · National culture

1 Introduction

As one of the leading industries of the new economy in the 21st century, the animation industry which is called the most promising sunrise industry, has significant influence on the global economy. In recent years, the animation industry in China has been booming, and the number and quality of animation have been improved to a certain degree. However, the animation industry in China is still in the preliminary stage. The level and capability of animation production are still a long distance compared with other animation powers such as Japan and the United States. At present, Chinese animation market is dominated by the United States and Japan while the Chinese animation is at the edge position [1].

However, Chinese animation had a glorious past. From the mid-1950s to the mid-1960s, the extremely difficult period, Chinese animators created a large number of classic works, such as the animated film "Uproar in Heaven", the ink animation "Baby Tadpoles Look for Their Mother" and "Herd Flute", which won awards throughout the world and have become world-renowned. They formally established the status of the "School of China" in the world animation industry. In the 1980s, ten years of turmoil ended, Chinese animation regained vigor and emerged animation works such as "Prince Ne Zha's Triumph Against Dragon King", "Three Monks", "Feelings of Mountains and Waters", "The Butterfly Spring" and "Nine-color Deer" which brought Chinese animation into the new splendor once again [2] (Figs. 1 and 2).

© Springer-Verlag GmbH Germany, part of Springer Nature 2020
Z. Pan et al. (Eds.): Transactions on Edutainment XVI, LNCS 11782, pp. 223–230, 2020.
https://doi.org/10.1007/978-3-662-61510-2_22

Fig. 1. Uproar in heaven

Fig. 2. Baby tadpoles look for their mother

We can create classical animation works in the 1950s, 1960s and 1980s with the extremely simple and crude funds and technical conditions. But today, with advanced technology, abundant funds, broad market, and the support of the state, Chinese animation is still stagnant with few high-quality products. What are the reasons? Apart from a few works such as "Lotus Lantern", "Monkey King and Ne Zha," few works have made a deep impression on the audience. Today Chinese animation appears to be in a trance, listless, and faltering steps because it lacks the spirit as it was originally and something of a spiritual nature. Therefore, in order to develop the Chinese animation, we must get back the soul of Chinese animation. Then what is the soul of Chinese animation? The answer goes to national culture [3] (Figs. 3 and 4).

Fig. 3. Herb flute

Fig. 4. Lotus lantern

2 Creation of Chinese Animation Scripts and Shaping of Chinese Animated Characters

2.1 Rich Materials for the Creation of Chinese Animation Scripts

With a long history, the extensive and profound national culture provides rich materials for the creation of Chinese animation scripts. The voluminous legends, folk tales, idiom allusions, literary works, drama scripts, and historical books are all inexhaustible treasury. It can be adapted based on the original, or can be taken a synopsis of the story and recreated. Even more, it can be selected the historical backgrounds, environmental facilities or story pieces as the material for another script creation. Taking China's first wide-screen animation film "Prince Ne Zha's Triumph Against Dragon King" as an example, the main storyline selected and based on the stories from 12th to 14th chapters of "The Canonization of the Gods" authored by Xu Zhonglin's of the Ming Dynasty in China. With rich content of the original book, the historical background is the period of prosperity of Shang and Zhou Dynasty. The storyline is King Wu's vassalage and attacks, which begins with "King Feng's visit to Nv Wa Palace" and end up with "King Wu's enfeoffment of dukes". There are a total of 100 chapters with numerous characters to show the brilliant scenes, such as "King Wen's visit to wise men", "Heng Ha's fight", "frozen in Mountain Qi", and "Huang Feihu's forcing five passes" (Figs. 5 and 6).

Fig. 5. Prince Ne Zha's Triumph Against Dragon King

Fig. 6. Prince Ne Zha's Triumph Against Dragon King

It is full of fancy and unique imagination. There are many spells such as flying and riding on clouds, changing shape and using suiton, summoning wind and rain, removing mountains and seas; powerful weapons such as god-hitting whip, god-killing sword, monster-revealing mirror, gold wand, silver-tip spear, steel spear, Hunyuan umbrella, sky-shaking seal, Qiankun steel ring; gods' residence such as Xiqi Chaoge, immortal mountains, seas and islands, triple realm palace, famous pass and town; beast rides such as Hualiu horse, leopard, white-forehead tiger, Chinese unicorn, david deer [4]. The animation "Prince Ne Zha's Triumph Against Dragon King" mainly selects the the stories from 12th to 14th chapters in Creation of the Gods. The creators keep the main storyline in the original work and delete the plots that are irrelevant or inconsistent to the creative theme, such as "Ne Zha's manslaughter to Lad Biyun", "Shifan's revenge", "Taiyi subdues Shifan", "Li Jing destroys the temple", "Ne Zha chases Li Jing" and so on.

The creators changed the enemy relationship between Ne Zha and his father Li Jing, turning the subject of the original story into a subordinate, and adding more warmhearted and reasonable content. The fight between Ne Zha and the Dragon King is adapted into the main body of the story. Flood and drought have a natural connection with the Dragon King. And up till now, some remote areas still have the ceremony of praying for rain. The character, Dragon King himself, is abhorrent, so it is more reasonable to make him the opposite of Ne Zha. It can also make the storyline more compact and better highlight the theme of the story. Meanwhile, the animation "Prince Ne Zha's Triumph Against Dragon King " retains the rich imagination in the originality including all kinds of spells such as flying to heaven and going into the sea, resurrection, stealth and so on, as well as various powerful weapons such as fire spear, Qiankun steel ring, Huntian ribbon, hot wheels etc. Moreover, the proper conversion of various scenes such as Lingxiao Temple, Crystal Palace, and Chentang Pass, all of which made the animation even more exciting and fascinating [5].

The successful experience of the animation "Prince Ne Zha's Triumph Against Dragon King" is worthy of our reference. It gives a satisfactory answer to make a better use of national culture. Simple copying cannot make a good script. It requires us to prune and adapt the original work, and then recreate according to the theme, and give the work contemporary characteristics, which can help absorb the rich nutrition of natural culture. Only in this way can they be used for our Chinese animation script creation and create excellent works that are favored by contemporary audiences [6].

2.2 Inexhaustible Role Images for the Shaping of Chinese Animated Characters

The profound historical and cultural accumulation produced a variety of vivid and numerous characters such as loyal chancellors and dutiful sons, sagacious emperors and good ministers, official ministers and force ministers, gentlemen and ladies, ancient gods, ghosts and monsters, pedlar and lackeys and so on. These artistic images left us abundant resources. On the one hand, these images are like the golds buried under the ground, which need us to explore, process, refine, and give them new life to shine in the new era. On the other hand, these images can provide many important references for us to create new characters. The creators can combine different characters and image features, according to the needs of their character shaping, to create new characters. They can also pursuit the arbitrariness and spiritual conformity like Chinese free-style painting. But the essence, energy and spirit of the character still carry a fresh and natural Chinese flavor.

The brave Chen Xiang in "Lotus Lantern" is adopted from the Chinese mythology "Hewing the Mountain to Rescue Mother". This character image has always been a classic representative of Chinese culture of filial piety. Dan Sheng, who always helps the weak and the poor in "Secrets of the Heavenly Book", comes from a character in the Ming dynasty novel "The Turn of The Flat Demon". All of these draw on the knowledge repositories of Chinese national culture [7].

"Three Monks" is one of the representative works in the history of Chinese animation. It won many awards of the world's film industry. Mr. Bao Lei adapts the film script based on a traditional Chinese folk saying "one boy is a boy, two boys half a boy, three boys no boy". It is directed by the late animation artists A Da and Ma Kexuan and the roles are designed by cartoonist Han Yu. The film contains the Chinese deep understanding of interpersonal relationship and tells the thought-provoking truth in an entertaining way. The three monks in the short film are all good people with shortcomings. They are selfish yet good-hearted. When designing the characters, Mr. Han Yu takes the inner heart of the characters as the revelation point, and carries out the character modeling with the vivid freehand brushwork technique of traditional Chinese painting. In the process of designing, he is not restricted by the realistic method of anatomy and proportion and conveys the spirit through form with the principle of "likeness and non-likeness".

The characters in "Three Monks" are both corny and clumsy. Under the comedy context, the childlike and lovable children's temperament and interest in the character modeling are highly highlighted. At the same time, clumsy and clever is the unity of opposites, both are inseparable. Besides the young monk and thin monk, the foolish-looking fat monk with oblate head and bloated body is clumsy yet skillful. The characters of the three monks are thin, young and fat. The colors are mainly red, yellow, blue and white.

At first glance, the crooked and twisted lines seemed to be drawn carelessly. The artistic conception of Chinese painting and the author's splendid painting technique can be understood only after you consider it carefully: To focus on the ideal rather than imitation and to focus on the memory rather than the intuition. By transforming exaggeratedly, the redundancy is removed, the essence is saved, a clearly defined primary and secondary is achieved, the priorities are balanced, the characteristics are presented. With "undecorated decoration", it creates a kind of simple beauty of simplicity [8].

In recent years, with the acceleration of globalization, some animations in the United States and Japan have learned experience from China. The image of the Monkey King in the Japanese cartoon "dragon ball" is based on the Chinese classic "Journey to the West". The "Mulan" of Disney Company in the United States, which is adapted according to the famous long poem "Mulan" in China. The characters in the film use many traditional Chinese ancient costumes and Chinese faces. The character in Disney style based on many Chinese elements is a great success. Chinese animation should do more beneficial searches in this area, looking for the integration point of national culture and modern aesthetic interest, and creating more vivid and lively animated images with strong national style and affluent connotation [9] (Figs. 7 and 8).

Fig. 7. Secrets of the heavenly book

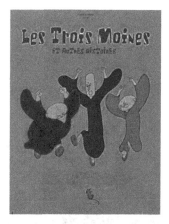

Fig. 8. Three monks

3   Importance of Chinese National Cultures for Animation

3.1   Giving Chinese Animation an Unique Charm

"Why is the water so clear in the dyke, for the fresh water comes from the springhead". Just like the Chinese poetry, Chinese national culture is the source of cultural connotation and artistic conception in Chinese animation. China is an ancient civilization with various cultures contending and blending here. The culture of Confucianism, Buddhist, Taoism and other cultures developed independently in this vast country, and they continued to interpenetrate and fuse, as a result of unique Chinese national culture. In the long river of Chinese history, the national culture gave birth to an artistic style with Chinese charm which contains the unique aesthetic pursuits and philosophical ideals of the Chinese people. This kind of charm is the essence of national culture. Only the essence of national culture is integrated into the animation, can we create outstanding animation works with more national and more worldwide characteristics.

Taking the animation "Feelings of Mountains and Waters" as an example, it is idyllic and full of sincere and moving emotions, revealing an ethereal and natural mood everywhere. The unique charm of Chinese culture makes this ink animation won the recognition of the not only Chinese audience, but also the worldwide [10]. The inspiration of "Thirty-six Chinese characters" comes from oracle. The art design fully reflects the simplicity of oracle, making the thousands of years' oracle jump vividly [11]. "The Butterfly Spring" absorbs the strengths of traditional painting and integrates the style of Yunnan decorative painting, which makes it unique in visual effect and impressive [12] (Figs. 9 and 10).

"Nine-color Deer" presents a perfect Dunhuang fresco art, it is the inheritance and development of Chinese Buddhist art and philosophy, creating a beautiful world of Zen. The audiences deeply feel a familiar and strong flavor when they see the film. It is different from the European and American realistic, logical and entertainment, it is also different from the Japanese animation that always make small things and feelings infinite amplification. It is a kind of feeling, artistic conception and the thick, dark,

Fig. 9. Feelings of mountains and waters

Fig. 10. Thirty-six Chinese characters

grand, magnificent and thought-provoking and philosophical Zen. It shows the spirit of the Chinese nation, traditional culture and ethnic flavor. The film absorbs the most profound spirit from the national culture [14].

3.2 Providing a Variety of Creative Forms for Chinese Animation

Traditional paintings, sculptures, puppets, and paper-cut shadows have enriched the expression forms of animation. And then different forms of animation have appeared, such as ink animation, paper-cut animation, origami animation, and puppet animation. In particular, ink animation has a unique Chinese charm, which has a huge contribution to the world animation. We should continue to explore and expand the forms of animation, such as clay sculpture animation, root carving animation, shadow animation, and ceramic animation, etc. Even more, we can also combine the various forms. Ink paintings, frescoes and new-year carvings provide materials for the modeling of animation figures. The creators should pay attention to absorb nutrients and extracting essences from Chinese classical arts, and then recreate so that they can meet the contemporary aesthetic tastes. For example, the model of general in the "The Pride of the General" is based on the face of the opera mask, the image of Monkey King in "Uproar in Heaven" is inspired by the model of folk prints and the opera mask. And the shepherd boy and buffalo in "Herb Flute" use Chinese ink painting. These works are filled with a strong national style.

In addition, background design, music effects, color use, scene processing and every aspect can draw nutrients from the national culture, so that Chinese animation is brilliant [15]. For example, the music and sound of the "Nine-color Deer" use a lot of traditional national Instruments. At the beginning of a film, the Chinese flute (Xiao) sound quickly show boundless desert to audiences. Pipa, Huqin and various kind of traditional percussion instruments create a mysterious remote western region character and style. An episode in the film, graceful and elegant, uses national vocal music singing performance to well describe the kindness and holiness image of Nine-color Deer.

4 Conclusion

To revitalize China's animation industry, we must inject the soul of Chinese national culture into Chinese animation. First of all, Chinese animators must fully realize the importance of Chinese national culture. Without the support of national culture, Chinese animation is just like a rootless tree. The blind pursuit of "take all", neglects our national culture, which is like climbing up a tree to catch fish. Always being along behind other countries is not a wise behaviour. Therefore, the nationalization of Chinese animation is the only way for its development.

What's more, we must continue to strengthen the study and accumulation of national culture. We should absorb nutrients not only from traditional Chinese painting, but also from various aspects such as literature, drama, and music, striving to create excellent animations with Chinese characteristics. Moreover, Chinese animation educators are required to reform the educational model, improve the heights of traditional cultural education, increase traditional cultural courses such as Chinese painting, appreciation of Chinese classics and Chinese history. It is necessary to develop more animation talents with profound national culture, which provides sufficient human resources for the development of national animation. Last but not the least, Chinese animation creators must actively practice, pursue innovation, and explore the new forms, new methods, new connotations and new styles of Chinese animation. To create a distinctive Chinese seal for Chinese animation and the world-class animation images like Monkey King and Ne Zha, and let Chinese animation fly again with the spirit and soul of the Chinese [16].

References

1. Tang, X.: A study of Chinese animation on the development of national cultural resources. Anhui University of Engineering (2016)
2. Hui, Y., Suo, Y.: The History of China's Animation Films. China Film Press, Beijing (2005)
3. Chen, H.: The new perspective on nationalization of Chinese animation creation. Movie Lit. **2018**(10), 117–119 (2018)
4. Xu, Z.: The Canonization of the Gods, pp. 12–14. The People's Literature Press, Chaoyangmen (1973)
5. Feng, J.: Upon the interpretation history of Creation of the Gods. Shandong University (2011)
6. Sun, J.: The cultural implication conveyed by the excellent traditional animation art—take "Prince Nezha's Triumph Against Dragon Kings" as an example. Art View **11**, 142–143 (2019)
7. Zhang, X.: From "Chinese school" to "Chinese style"—a comprehensive view of the development of domestic animation in 70 years. Art Wide Angle **05**, 15–25 (2019)
8. Li, R.: The "nationalization" study of modeling and performance-animation character design. Zhejiang Sci-Tech University (2017)
9. Wang, N.: A study on the re-modeling of Chinese traditional artistic images in animation creation. Northwestern University (2013)
10. Zhang, B., Huang, Y.: An analysis of the symbolic characterization of "Chinese School" ink animation–taking "Herd flute" and "Feelings of Mountains and Waters" as an example. Art Stud. **04**, 66–68 (2019)
11. Woo, Y.: A study on the evolution of national nature of national animation in new china in 70 years. Film Rev. **19**, 47–52 (2019)
12. Dai, P.: Interpretation of chinese classic cartoon <butterfly spring>. Admire **36**, 173–174 (2012)
13. Cai-rang, M.: Study of <34 Buddha's life-Nine color deer life>. Qinghai Normal University (2019)
14. Zhou, L., Yang, W.: An analysis of the artistic features of the cartoon "nine color deer". J. Jiangxi Norm. Univ. Sci. Technol. **05**, 118–120 (2010)
15. Chang, G.: Take out more excellent cartoons with national and cultural characteristics. N. Film (01) (2006)
16. Chang, G.: Globalization and the thinking of Chinese animation. Film Lit. **17**, 37–38 (2010)

Animation and Miscellaneous

Fruit Shape 3D Printing Based on Wavelet Interpolation Modeling

Zhifa Du, Tianlong Yang, and Tonglin Zhu[⊠]

Institution of Agricultural Multimedia Technology, South China Agricultural University,
Guangzhou 510642, Guangdong, China
1344100133@qq.com

Abstract. In order to build a large model database for fruit 3D printing and require
(1) sampling data is as simple as possible; (2) data volume and data structure are
unified; (3) modeling method has multi-resolution capabilities, this paper presents
a fruit shape modeling method based on wavelet interpolation, trying to uniformly
sample 2 to the power of m longitude lines of fruits with a upright axis, and
uniformly sample 2 to the power of n points on each longitude line. Then, we use
wavelet interpolation to obtain the longitude and latitude lines models that meet
the required precision for 3D printing. Additionally, we discuss several special
issues of fruit shape 3D printing and Gcode file generation. The experimental
results show that the modeling method has the advantages of good model effect,
short printing time, and saving of printing material, which can achieve simple and
practical application requirements.

Keywords: Wavelet · Interpolation · Modeling · 3D printing · Internal supports

1 Introduction

Based on the digital model file, 3D printing uses the binding material such as powder
metal or plastic to construct the objects through layer-by-layer printing. With the important value in scientific research, teaching and entertainment, the fruit 3D printing can be
applied to printed foods required by astronauts' psychology and to the agricultural products displayed in museums and exhibition halls. In recent years, scholars have studied
various algorithms to establish mathematical models for fruits. For example, Seah et al.
used the modeling approach of Ball B-Spline function to represent the tubular and crust-
like three-dimensional objects in free form [1]; Qiao et al. put forward the NURBS-based
plant surface modeling measurement which uses the three-dimensional scanner to obtain
control points varying from 100 to 300 on the surface of cucumber, pepper, and corn
and then adjusts the weight and density of the control points to construct their models
[2]; Wang et al. used the Bezier curve to model the Longan berries and simulated the
forms of different growth processes of the Longan berries by adjusting the central axis
base points, control points and section feature points [3]; Yuan et al. used the B-Spline
curve surface to fit the tomato shape. After obtaining the three-dimensional coordinates
of the points on several curves of the fruit surface, they used uniform spline curves to

© Springer-Verlag GmbH Germany, part of Springer Nature 2020
Z. Pan et al. (Eds.): Transactions on Edutainment XVI, LNCS 11782, pp. 233–245, 2020.
https://doi.org/10.1007/978-3-662-61510-2_23

fit these curves and used spline surface to simulate the three-dimensional shape of the tomato fruit [4]. The fruit modeling methods mentioned above have non-uniform sampling data volume and data structure and every change of the resolution of the model needs to recalculate all points, that is, have no function of multi-resolution analysis. In addition, the traditional 3D printing process uses the model built by the above methods to first be converted into a triangular plane before getting the intersecting line and being converted into Gcode command, which will result in multiple conversion processes and large amount of calculation [5], therefore they cannot adapt to the 3D printing well. Therefore, there must have the new modeling approaches to model with latitude directly before printing the latitude and the method of wavelet interpolation can just adapt to the latitude modeling well. The key for the modeling lies in simple sampling. The simplest method is to uniformly sample the height and then uniformly sample the latitude. This paper tries to take 2^m longitudes by the uniform sampling method from the fruit with upright axis and take 2^n points by the uniform sampling method from each longitude. Then this paper uses interval wavelet first for the uniform interpolation of the longitudes and then uses the periodic wavelet for the uniform interpolation of the latitude line data obtained from the last interpolation. The interpolation principle of periodic wavelet can ensure the smooth connection of end to end of the discrete sampling periodic wavelet after the interpolation and the interpolation error for the ellipse and the circle is small. There is no way to do a good job in the two points mentioned above if the uniform spline curve is used for the modeling. What's more, the wavelet interpolation method has multiresolution feature. When the resolution is improved, the value of the previous layer can be used to calculate the interpolation of the next layer, which can greatly reduce the calculation amount.

In the research of saving materials of 3D printing, the scholars have also proposed various algorithms to reduce the cost of printing materials in recent years. For example, Li et al. put forward the internal support lightweight method of density perception [6] and Fu et al. proposed the material saving optimization method [7]. In terms of the spatial characteristics of fruits with upright central axis, this paper puts forward a new method of material saving which can save more materials than the above methods.

2 Algorithms and Steps

2.1 Data Collection

The fruits with the upright central axis are divided into three categories, namely: (1) ones with sunken upper and lower sides, such as apples, persimmons, etc.; (2) ones with sunken lower side, such as Sydney, jujube, etc.; (3) ones with no upper and lower sides, such as orange, grapefruit, longan berries, grapes and so on and only the fruits with sunken upper side can be included in category 2.

According to the above categories, the data of the first category of fruit is collected with the aid of a 3D digitizer. It is convenient to obtain the three-dimensional coordinates of any point on the surface taking the center of the bottom of the fruit as the origin coordinate and to obtain its eight longitude lines. One of the longitude lines is taken as the case to be divided as shown in the Fig. 1.

Generally, the sunken depth of the upper and lower side of the fruit will not exceed 1/3 of the height of the whole fruit. Thus, as shown in Fig. 1, the whole height of the

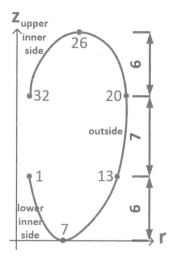

Fig. 1. Longitude line division map (the abscissa is the distance r from the point on the longitude line to the fruit axis, and the ordinate is the height z)

fruit is divided into three parts with the ratio of 6:7:6. The sampling points of the whole longitude starts from the lower inner side and ends at the upper inner side. The point sampling is carried out according this ratio and order and there are 32 points as a total. If the sunken depth of the fruit does not reach 6/19 of the overall height then the sampling points of the inner longitude lines can be reduced, that is, the sampling of the 1st, 2nd, and 32nd, 31st, … points is omitted.

As the curve of the longitude is not an injective function to be unable to perform the interpolation hence the method used here is to take the horizontal plane where the bottom and topmost points of the longitude line are in as the interface and to have mirror-imaging on the inner two section of longitude lines. Figure 2 is a real longitude line of the sample of the first category of fruit (the abscissa is z and the ordinate is r) and Fig. 3 is the result of the mirror-imaging. The sunken depth of the upper and lower sides of this longitude line did not reach the 6/19 of the overall height thus the number of sampling points is less than 32. (In fact, there are only 28 points, missing 3 points in the lower side and missing 1 point in the upper side).

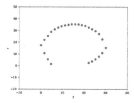

Fig. 2. The shape of a longitude line

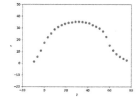

Fig. 3. The shape of the longitude line after mirror-imaging

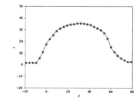

Fig. 4. Radius function of the longitude line of the first category of fruit

In addition, because the wavelet interpolation algorithm requires 2^n points for the calculation, in order to satisfy this condition, the r values of the last sample points at both ends can be used to fill the missing sample points at both ends to complement the 32 points and meet the requirements of the points. The results obtained by this method are shown in Fig. 4 (only dots) and these 32 points are called the radius function of the longitude line.

Similarly, the sampling and processing methods for the longitude lines of the second and third category of fruit are similar to those of the first category and the radius functions of a longitude line of the second and third category of fruit are shown in Figs. 5 and 6 (only dots).

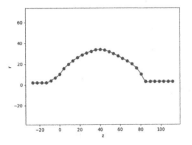

Fig. 5. Radius function of the longitude line of the second category of fruit.　　**Fig. 6.** Radius function of the longitude line of the third category of fruit

2.2　Interpolation of Longitude and Latitude Lines

The wavelets used in this paper are mainly periodization of Daubechies compact support wavelets [8] and Daubechies interval wavelets [9] derived from Daubechies compact support wavelets. According to the two-scale equation and the construction method of periodic wavelets given by literatures [10, 11], supposing that functions $f(x) \in L^2[0, 2\pi]$, $f(x)$ have 2^j uniform sampling points in $[0, 2\pi]$, that is $f^j\left(\frac{2\pi k}{2^j}\right)$ ($k = 1, 2, \ldots, 2^j$), then

$$c_k^j = \langle f^j, \varphi_{jk}^{per} \rangle \tag{1}$$

$$d_k^j = \langle f^j, \psi_{jk}^{per} \rangle \tag{2}$$

($k = 1, 2, \ldots, 2^j$)

$$c_k^{j+1} = \sum_k [h_{n-2k} c_k^j + g_{n-2k} d_k^j] \tag{3}$$

$$f^{j+1}(x) = \sum_{k=2}^{2^{j+1}-1} c_k^{j+1} \varphi_k^{j+1}(x) \tag{4}$$

$$d_k^{j+1} =< f^{j+1}(x), \psi_k^{j+1}(x) > \tag{5}$$

This is the method to have interpolation on the periodic function $f(x) \in L^2[0, 2\pi]$ using periodic wavelets, that is, the scale function coefficient c_k^j and wavelet function coefficient d_k^j of the same layer are calculated from the value of f^j, as well as the Eqs. (1) and (2), and then followed by the scale function coefficients c_k^{j+1} of the higher level calculated by Eq. (3), the function interpolation $f^{j+1}(x)$ of the higher level calculated by Eq. (4) and wavelet coefficient d_k^{j+1} of the higher level calculated by Eq. (5), giving the way to an infinite loop.

The transformation of the Daubechies interval wavelets can also be carried out according to the method in the literature [10, 11].

Supposing that f(x) is a function to be interpolated and its length is 1. It has sample value f_k^j (k = 1, 2,...,) at $x = \frac{1}{2^j}, \frac{2}{2^j}, \ldots, \frac{2^j}{2^j}$. Similar to the Daubechies periodic wavelet algorithm Eqs. (1)–(5), we can calculate the coefficients c_k^j, d_k^j of each layer.

$$c_k^j =< f^j, \varphi_k^j > \tag{6}$$

$$d_k^j =< f^j, \psi_k^j > \tag{7}$$

$$c_k^{j+1} = \sum_k [h_{n-2k} c_k^j + g_{n-2k} d_k^j] \tag{8}$$

$$f^{j+1}(x) = \sum_{k=0}^{2^{j+1}-1} c_k^{j+1} \varphi_k^{j+1}(x) \tag{9}$$

$$d_k^{j+1} =< f^{j+1}, \psi_k^{j+1} > \tag{10}$$

And then we can calculate f^{j+2} and so on with this algorithm.

Taking the first category of fruit as an example, the radius function of the longitude line of the fruit can be interpolated with the interval wavelet interpolation principle, that is, the value of the arbitrary precision of the radius function can be obtained with the Eqs. (6)–(10), and the interpolation result is as shown in the curve of Fig. 4.

Finally, we remove the curve segments that are extended by extra padding points, and according to the same mirror-imaging method, with the z-value of the lowest and top-most points of the original longitude line as the boundary, the radius function of the longitude line can be converted back to the shape of the original longitude line. In this way, the values of any precision of each longitude line can be obtained. Figure 7 shows all eight longitude lines of the first category of fruit obtained after the above processing.

It is obvious that the converting from mirror-imaging step of the longitude line from radius function will produce a tiny spike at the lowest and topmost part of the longitude line. This problem will be resolved in Sect. 2.3.

After getting fine longitude lines, we can obtain the contours described by 8 points at any cut-layer. Here the radius function of a closed curve is introduced and a closed curve can be written as the polar coordinates equation r = r(θ) with its center as the origin point and the distance from the point on the curve to the origin point as the radius.

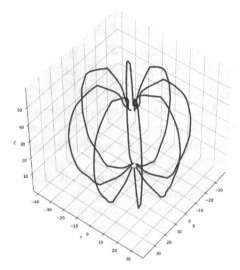

Fig. 7. All the longitude lines of the first category of fruit

Under the cartesian coordinate system, the figure drawn with as the θ and r is the radius function of the closed curve [12, 13]. Taking the closed curve of a certain cut-layer, its radius function is shown in Fig. 8 (dot only, the abscissa is θ, and the ordinate is r). With use of periodic wavelet interpolation method and 2π as the period, values of arbitrary precision of the radius function of that cut-layer curve can be obtained by using the Eqs. (1)–(5), as shown in the curve of Fig. 8.

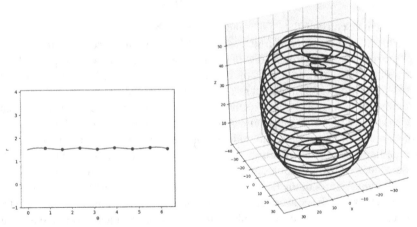

Fig. 8. Radius function of the latitude line   **Fig. 9.** Latitude lines of all the cut-layers (schematic diagram only)

Then the radius function can be converted into a closed curve of the original latitude line with $\begin{cases} x = r\cos(\theta) \\ y = r\sin(\theta) \end{cases}$ and in this way, the fine latitude lines of all cut-layers can be obtained, as shown in Fig. 9.

Similarly, the processing method of the longitude line of the second and third category of fruit and the interpolation of the latitude lines can be obtained as described above, and the results are shown in Fig. 10 and 11.

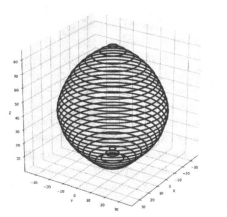

Fig. 10. All latitude lines of the second category of fruit

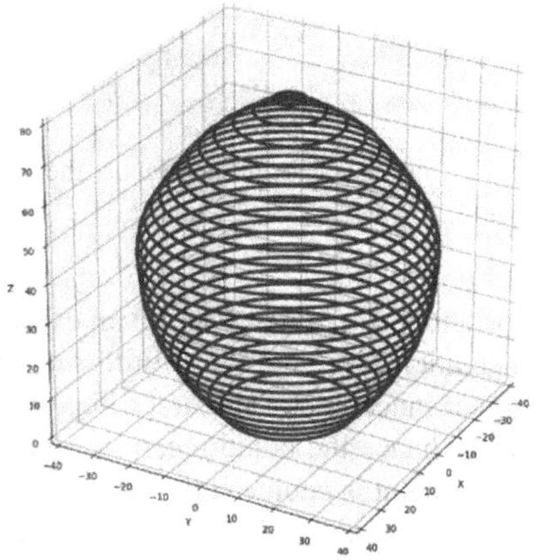

Fig. 11. All latitude lines of the third category of fruit

It can be seen that according to the division method of Fig. 1, the specifications of the sampling data of the three categories of fruits can be unified.

2.3 Special Treatment

First of all, as the 3D printer (Qubea QD-1x) used in this experiment is an FDM-based printer and the first cut-layer of the printed object must have a certain size of contact surface with the printing platform. Or, the contact surface is too small, the object is difficult to stick to the platform, resulting in failed printing. Therefore, the treatment is to cut off the fruit model with the plane (such as z = 1.5 mm) with small distance from xoy plane and this treatment is achieved by deleting part of latitude data at the bottom. After this treatment, the bottom surface of the fruit is a ring belt instead of the original ring-shaped thin line, giving a large enough contact surface with the printing platform and solving the problem of tiny spike in the lower part after the mirror-imaging of the longitude in Sect. 2.2. The simulation printing effect of the lower part of the first category of fruit is shown in Fig. 12:

Secondly, the upper surface of the first category of fruit is suspended, hence the direct printing will cause collapse. The traditional solution is to fill the inside, that is, to print the interior of the fruit in solid form, which will cause the great waste of printing materials and time consuming. However, noticed that all the fruits studied here have

Fig. 12. Simulation printing effect at the bottom

Fig. 13. Simulation printing effect of the generated column

an upright central axis therefore an extra middle column can be printed on the original mathematical model of the fruit to support the upper surface. From the latitude line in the upper inner side of the first category of fruit, it is easy to know the span of the first cut-layer hance the column can be added according to the span size of this cut-layer. This treatment is achieved through adding additional Gcode command and this will be described in detail in Sect. 2.4. The simulation printing effect is shown in Fig. 13.

Thirdly, in Sect. 2.2, there is a tiny spike in the upper part of model of the first category of fruit, which can be expressed as that the upper inner side is getting closer to the outer side of the latitude line and they form a thin line at the joint finally in the actual printing. The solution is to stop printing the latitude line on both sides when the upper inner side latitude line is close enough to the outer side latitude line, but to start to print a horizontal band between them, which can fill the gap. And then in the subsequent layer, the band gradually narrows to a certain extent then stop printing. This treatment is achieved through modifying the Gcode command and will be described in detail in Sect. 2.4.

2.4 Gcode Generation

As a control command file that guides how the print head of the 3D printer moves, the Gcode file is a kind of plain text and one command can be written in one line. The traditional 3D printing process is converting CAD model to the STL file, that is, the surface is represented as a file including several triangular planes and then converting the STL file to Gcode file. Based on the preset layer thickness, in this process, several horizontal planes with the layer thickness as the interval intersects the triangular planes model to obtain all the intersection lines. These discrete points on these intersection lines are points at which the printhead will traverse and the 3D printing can be finished by extruding the printing material between these points.

The fruit modeling process in this paper has obtained accurate latitude line data for each of the three categories of fruit. In fact, the function of latitude line data from each layer is equivalent to the intersection lines mentioned above. Therefore, the direct conversion of the latitude line data into the printhead control command, that is, a Gcode file will finish the conversion process of the last step of 3D printing.

In the Gcode file, the commands that control the movement of the printhead and the printing material extrusion are mainly G1 and other commands (the format of the G1 command is G1 Xnnn Ynnn Znnn Ennn Fnnn, where nnn represents the number, the first three parameters represent the coordinate of the printhead movement, the E parameter indicates the printing material extrusion amount, and the F parameter indicates the moving speed). According to the format of the Gcode command, the sequence of points of each layer in the fruit model can be converted into a series of Gcode commands. The pseudo code of the Gcode command generated for the first category of fruit is as follows, where the following "generate gcode" operation means the generation of a series of G1 commands to control the printhead to move and extrude the printing materials according to various curves:

```
for current_outer_latitude_line in all_outer_latitude_lines:
    if current_layer_number < preset_solid_printing_leyer_number:
        generate_gcode(solid_print_between_inner_and_outer_latitude)
else:
    generate_gcode(print_outer_latitude)
    if inner_latitude_exists_in_current_layer:
        if distance_between_inner_and_outer_latitude > fixed_value_1:
            generate_gcode(print_inner_latitude)
        else:
            generate_gcode(print_gradually_narrowing_belts)
            break# ignoring remaining layers that cause spike
             # At this stage, the inner latitude appears, stop printing column
        clear_the_outline_record_of_the_column
    if outline_of_column_is_recorded:
    generate_gcode(print_column)# At this stage, 3 curves are printing simultaneously
    if maximum_radius_of_latitude <=fixed_value_2:
        record_outline_of_column
z+=0.1# printing head moves up by 0.1 mm
```

Therefore, the above conversion algorithm can convert the mathematical model of the first category of fruit into specific printhead movements route and print out the first category of fruit.

The logic for the second and third category of fruit is simpler because there is no need to print the column and handle the spike at the top.

3   Experimental Results and Analysis

With the Python3 and Qubea QD-1x 3D printers (as shown in Fig. 14) as the experimental platform, the longitude lines of the fruit are divided according to rules mentioned before and 8 (2^m) longitude lines is uniformly sampled with 32 (2^n) points uniformly sampled on each line, which makes that the sampling of the fruit with upright axes has uniform data volume and data format, that is, the height of the fruit H (in mm), the number of longitude line sampled $M = 2^m$ (each line is separated by $2\pi/M$), the number of sampling points per line $N = 2^n$, and the coordinate data of M*N sampling points $\{(x_i^j, y_i^j, z_i^j)_{i=1}^N\}_{j=1}^M$ (unit as mm). According to the interval wavelet interpolation principle, the 8 longitude lines of the three categories of fruit are uniformly interpolated from 2^5 points to 2^{12} points on the z-axis and then with a period of 2π, the 8 points of each layer are uniformly interpolated to 2^{12} points according to the periodic wavelet interpolation principle. The mathematical model obtained will be converted into a Gcode file to perform the layer-by-layer printing based on the latitude line. This treatment method to convert the model into Gcode file directly also breaks the traditional 3D printing process, reduces the error caused by multiple conversions and improves the smoothness of printing of latitude line. The time that the original data is generated to the Gcode file has saved because the steps of generating the triangular planes have been skipped. For example, when there is the necessary to double the resolution of the print model, the traditional 3D printing process needs to regenerate the triangle planes before calculating intersection lines, and Gcode file conversion. This kind of repetitive computation will take time at least 4 times than before while the method in this paper can continue to interpolate on the basis of the original to get new model points directly. Therefore the treatment time for this method will not change.

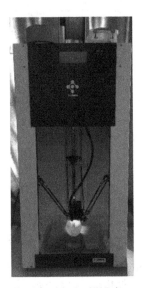

Fig. 14. QD-1x 3D Printer

Figure 15 is the effect of the first category of fruit just beginning to print the upper surface. It can be seen that the center column supports the upper surface very precisely for the correct printing, thereby achieving the print of support part only where necessary and avoiding unnecessary overall internal filling. Compared to the traditional printing methods, the method of this paper significantly saves printing materials and saves lots of printing waiting time.

Figure 16 is the printing effect diagram for the uncorrected top spike and Fig. 17 shows the effect of corrected top spike. Comparing the Fig. 16 with Fig. 17, it is not easy to see the spike for the first category of fruit, making the printing effect more realistic.

The final printing results for the three categories of fruits are shown in Figs. 18, 19, and 20.

Fig. 15. Printing effect diagram with column as the support

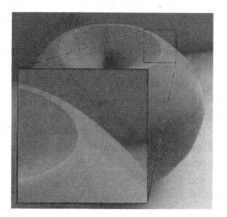

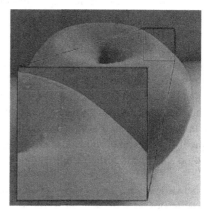

Fig. 16. Printing effect diagram for top with spike

Fig. 17. Printing effect diagram for one with spike removed

Fig. 18. Apple

Fig. 19. Snow pear

Fig. 20. Orange

As described in Sect. 2.3, the bottom of the fruit is cut off for the physical require-ments of printing and it is difficult to perceive the cutoff height which is very tiny to be less than 2 mm.

4 Conclusions

To make the conclusion of the above fruit modeling and 3D printing process, this paper has the following two innovations:

(1) The Daubechies interval wavelets and periodic wavelets are used to interpolate the longitude and latitude lines of the fruit respectively to make the fruit modeling have the multi-resolution function, that is, there is no need to repeatedly calculate all the points but just increase the interpolation points on the original basis when the model is enlarged, which can reduce calculation greatly.

(2) The direct conversion algorithm of the longitude and latitude line model of the fruit to the Gcode file is achieved, which can reduce the error in the model conversion process and improve the conversion speed.

In conclusion, the fruit shape model based on wavelet interpolation can meet the data requirements of 3D printing well and the effect of the method in this paper has achieved the practical requirements to be able to build shape model library of fruit with upright axis in large scale. But the effect of 3D printing may be somewhat distorted when there is the need to print fruit with very large size or the fruit with larger radius of curvature of the upper surface. In this case, the method of this paper needs further improvement. For some fruits with an upright axis and complex shapes (such as carambola, pineapple), the method also needs further improvement and modification.

Acknowledgments. This work is supported by Research Fund for the Doctoral Program of Higher Education of China. No. 2012440410018.

References

1. Seah, H.S., Wu, Z.K.: Ball B-Spline based geometric models in distributed virtual environ-ments (2005)
2. Qiao, G.: Modeling and application of plant surface modeling based on NURBS. Dalian University of Technology (2010)
3. Weibin, W., Liyu, T., Chongcheng, C., et al.: Simulation study of geometric modeling and growth process of Longan berries. J. Agric. Sci. Technol. **03**, 84–91 (2010)
4. Yuan, X.: Research on three-dimensional shape modeling and mesh model optimization method of tomato. CNU(Capital Normal University) (2012)
5. Cai, J., Li, W., Liu, J.: All-in-one 3D Printing. Tsinghua University Press (2016)
6. Dawei, L., Ning, D., Xiaotong, J., et al.: Density aware internal supporting structure light-weighting modeling of 3D printed objects. J. Comput. Aided Des. Comput. Graph. **05**, 841–848 (2016)
7. Chilin, F., Bin, L.: Research on 3D printing materials saving optimization related technology. J. Comput. Aided Des. Comput. Graph. **04**, 742–750 (2017)
8. Cohen, A., Daubechies, I., Vial, P.: Wavelets on the interval and fast wavelet transforms. Appl. Comput. Harmonic Anal. **1**(1), 54–81 (1993)

9. Daubechies, I.: Ten lectures on wavelets. Siam (1992)
10. Boggess, A., Narcowich, F.J.: A First Course in Wavelets with Fourier Analysis. Wiley, Hoboken (2015)
11. Zhang, G., Zhang, W., Xue, P.: Wavelet analysis and application basis. Northwestern Polytechnical University Press Co. Ltd. (2006)
12. Zeng, Q., Zhu, T., Zhuang, X., et al.: Periodic wavelet descriptor of plant leaf and its application in botany. Int. J. Wavelets Multiresolut. Inf. Process. **13**(06), 20–41 (2015)
13. Qingmao, Z., Huiqin, L., Tonglin, Z.: Geometrically modeling leaf outline based on wavelet descriptors. J. Comput. Aided Des. Comput. Graph. **12**, 2046–2053 (2011)

Research on Digital Forensic Investigation of Xbox One

Ying Zhang[1], Feng Gao[1], and Hong Guo[2]($\boxtimes$)

[1] The Third Research Institute of Ministry of Public Security, Shanghai 201204, China
{zhangying,gaofeng}@stars.org.cn
[2] Academy of Forensic Science, Shanghai 200063, China
guoh@ssfjd.cn

Abstract. Since the first video game console called Pong was released in the 1970s, game consoles have grown continuously with the economic and technological development. There are some leading gaming manufacturers in current market, including Sony, Microsoft and Nintendo. As the eighth generation of video game console published by Microsoft, Xbox one is one of the most popular game consoles and thus attracts numerous customers. It is noticeable, however, Xbox one can no longer be treated as consoles but rather as computer-like machines. In this paper we are going to analyze partition, file system, relevant file and other data of Xbox One concretely.

Keywords: Xbox one · Digital forensic · Video game console · File system

1 Introduction

Xbox one was published in 2013 and treated as the successor to Xbox 360 which was introduced in 2005 as part of the seventh generation of video game consoles [1]. Different from its predecessor, Xbox one switches from PowerPC-based architecture to x86, which was utilized in the original one. Some accessories were redesigned on the basis of Xbox 360 as well, including controller, D-pad and triggers with the ability of delivering directional haptic feedback. Besides that, Xbox one turned to focus on cloud computing, social network, video sharing and recording with emergence of new concepts of Internet. Due to above new features and improvement, Xbox one has been broadly welcomed since its release and attracted numerous gamers. According to research from firm HIS Markit, Microsoft's Xbox one family of consoles sold 39.1 million units globally by the end of March 2018 [2].

It is well known that Xbox one is not only a game console but also computer-like machines. Concretely speaking, it owns its unique operating system, file system, partition and system file. Previous research has shown that this console has been utilized in criminal activities including identity theft, child pornography, economic fraud and so on [3]. Thus more and more investigators have put their emphasis on Xbox one forensics.

© Springer-Verlag GmbH Germany, part of Springer Nature 2020
Z. Pan et al. (Eds.): Transactions on Edutainment XVI, LNCS 11782, pp. 246–255, 2020.
https://doi.org/10.1007/978-3-662-61510-2_24

1.1 Background and Research Review

Reviewing the progress of Xbox forensics, different research have been conducted regarding Xbox family. For example, Podhradsky and D'Ovidio [4] mentioned an approach that explores Xbox 360 preliminarily. Researchers concentrate on analyzing partition, system update files, personal data like username and password, messages and related data via using professional tools. Fortunately there are some special tools for Xbox 360 such as Xplorer360, which is helpful for Xbox 360 investigation. Using these tools, investigators could examine, analyze, search and signify relevant data easily. Bolt [5] demonstrated concrete digital forensic process and correlative technology, analyzed technology that was being utilized at present, and pointed out the deficiency of current investigation. Starting with analysis of hardware and configuration, he described Xbox 360 concretely and gave a further explanation of cache folder, content folder, mindex folder and so on.

As for Xbox one [3], previous research has shown the partitions, file metadata and special system file as well. Distinguished from Xbox 360, it is not that easy to interpret content and structure of system files due to lack of proper tools. Nevertheless, Steven utilized a new way to determine meaning of each system file, while entropy score was introduced to explore the nature of each file. The entropy score was expressed as the minimum number of bits needed to be encode each byte of information in the most optimal compression regime as mentioned in previous research [6].

1.2 Technical Issues

However, investigation towards Xbox one has been an issue for a long time for various reasons. First of all, different from Windows or Linux, the operating system is difficult to analyze without regular pattern. It is believed that this system was customized for Xbox one only. Secondly, there are special system files in this game console which are difficult to analyze and examine. In addition, it is difficult to find professional tools with high pertinency such as Xplorer360. All those difficulties are supposed to be solved in the future work.

For the scope of this paper we will conduct various operations to a brand new Xbox one, including factory settings restore, network log on, user log on, system update, game installation and game play. Section 2 will describe above operations concretely. After each step, hard drive image will be conducted and recorded towards corresponding operation. Section 3 will describe partition layout, file system, system data and so on. More analysis of system files will be given in this chapter as well. Conclusion is in Sect. 4.

2 Methodology and Implementation

In order to give some insights into partitions, systems and data structures of Xbox one deeply, we will perform a sequence of actions to the console. The point is that hard drive should be imaged after each step to record data via using write-block device. After that, data contained in hard drive will be compared so as to find differences between various stages and analyze value of each file.

2.1 Methodology

As mentioned above, various operations are likely to be conducted to the game console as shown in Fig. 1. Details of each operations will be described concretely as follow.

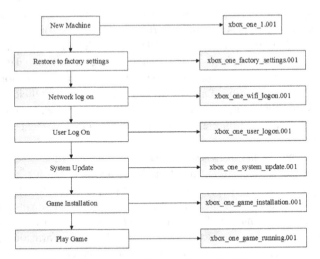

Fig. 1. Various operations performed to Xbox one

New Machine. One new Xbox one was purchased for this experiment and its hard drive will be removed from the machine. Via using write-block device the whole disk will be imaged. Moreover, the image is required to be hashed to implement date integrity and validity. After that, the drive is supposed to be installed back to machine.

Restore to Factory Settings. After disk installation, the machine will be turned on. Through browsing user interface, we select 'reset console' under 'configuration' so that the game console would be restored to factory settings. Image is required to be created as well during this step.

Network Log On. For a machine restored to factory settings, the first thing is to log on the network. After turning on the console, we select 'wireless network configuration' and then log on the target network via using user name and password.

User Log On. As mentioned in previous chapter, Xbox one started to focus on the capabilities of social communication, while Xbox Live is designed to realize this function. User account is supposed to be created and configured for Xbox live. Besides that, system activation for new machine requires user log on as well. User information could be recorded while selecting 'log on'.

System Update. After user log on, system update is forced to conduct. No further actions could be taken until the update is complete. Operations could be conduct via following steps displayed in the screen.

Game Installation. Game installation is an important part of the whole inspection. Data files of game will bring us new clues and inspirations to data analysis. Fallout 4 would be selected and installed on this game console. In addition, game update is required as well after installation.

Play Game. Fallout 4 will be played after installation and update. When starting running, some basic elements such as character's appearance, name and attributes is required to be configured to start the game. During the process, it is necessary to save game at different point and go through various scenes, so that more evidences will be record on the hard disk.

2.2   Implementation

All those steps discussed above are supposed to be implemented through specific equipment and tool. Next we are going to demonstrate details and purpose of different tools.

Hard Drive Image Obtaining. There are several ways to image hard disk in game console. Generally speaking, forensic investigators prefer professional tools such as X-Ways Forensics and Encase, or equipment like Forensic Falcon. The difference is that X-Ways Forensics or Encase is software only while Forensic Falcon includes associated hardware like write-blocker as well.

Forensic Falcon is a forensic imaging solution available produced by Logicube, achieving speeds of over 30 GB/min [7]. We use it in our experiment due to its abilities of write-block. Forensic Falcon was turned on first, and then the hard drive took from game console would be connected to the equipment via using SATA wire. Point is that read-only interface is required to be utilized. Image acquisition will be started after configuration of relevant parameter such as source, destination and way of image. The process of making a hard disk image is shown as Fig. 2. Once finished, we will acquire an image with extension ".001".

Data View, Search and Analysis. After image acquisition, we turn our attention to how to read image file as well as view, search and interpret files from image. Most traditional forensic tools can achieve that goal.

In our experiment, X-Ways Forensics will be utilized to interpret data from image. It is know that X-Ways Forensics is an advanced work environment for computer forensic examiners [8]. It has the ability of reading partitioning and file system structures inside raw image files, automatic identifications of lost or deleted partitions and so on. Hence information like partition and file system would be demonstrated via using this tool.

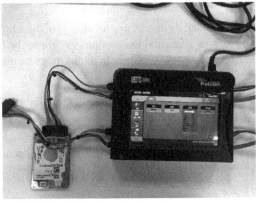

Fig. 2. Image the hard drive via using Falcon

Hash Calculation. After obtaining data file from each partition, we are going to compare modifications for the same file between various stages. The simplest and most effective way to find the difference is calculating hashes of files with the same name and comparing them.

In our experiment, Karen's Hasher will be utilized to calculate SHA256 for each file. Karen's Hasher is a small utility to calculate the hashes of a file, a list of files or text [9]. It is known that as the SHA256 of one specific file modifies, and thus the file differently.

3 Data Analysis

Via viewing and analyzing the images created in the above operations, it is found that there are a number of commonalities in aspect of partition data structure between different stages. We are going to demonstrate those details and give further analysis and research in the following chapters.

3.1 Partition Layout

As interpreted from image, the hard disk of game console is composed of five partitions and one unpartitioned space. The details of partitions are illustrated in Table 1.

Table 1. Partition layout for Xbox one.

Name	File system	Total capacity	Sector count
Temp content	NTFS	41.0 GB	85983232
User content	NTFS	365 GB	765460480
System support	NTFS	40.0 GB	83886080
System update	NTFS	12.0 GB	25165824
System update 2	NTFS	7.0 GB	14680064
Unpartitioned space	GPT		

According to our research, there are no modifications of partitions' name and number in any stage. In other words, it is believed that each hard disk of Xbox one is consist of those 5 NTFS partitions.

3.2 Temp Content

After examining the files under this partition, we find that it is difficult to determine the significance and use of this partition. Figure 3 illustrates files located under temp content in the stage of restore to factory settings.

Name ▲	Type	Path	Size	Created	Modified	Record changed	Accessed
$Extend		\	9.0 MB	2014/09/02 00:00:06	2014/09/02 00:00:06	2014/09/02 00:00:06	2014/09/02 00:00:06
(Root directory)				2014/09/02 00:00:06	2014/09/02 00:00:50	2014/09/02 00:00:50	2014/09/02 00:00:50
$AttrDef		\	2.5 KB	2014/09/02 00:00:06	2014/09/02 00:00:06	2014/09/02 00:00:06	2014/09/02 00:00:06
$BadClus		\	0 B	2014/09/02 00:00:06	2014/09/02 00:00:06	2014/09/02 00:00:06	2014/09/02 00:00:06
$Bitmap		\	1.3 MB	2014/09/02 00:00:06	2014/09/02 00:00:06	2014/09/02 00:00:06	2014/09/02 00:00:06
$Boot		\	8.0 KB	2014/09/02 00:00:06	2014/09/02 00:00:06	2014/09/02 00:00:06	2014/09/02 00:00:06
$LogFile		\	64.0 MB	2014/09/02 00:00:06	2014/09/02 00:00:06	2014/09/02 00:00:06	2014/09/02 00:00:06
$MFT		\	256 KB	2014/09/02 00:00:06	2014/09/02 00:00:06	2014/09/02 00:00:06	2014/09/02 00:00:06
$MFTMirr		\	4.0 KB	2014/09/02 00:00:06	2014/09/02 00:00:06	2014/09/02 00:00:06	2014/09/02 00:00:06
$Secure		\	0 B	2014/09/02 00:00:06	2014/09/02 00:00:06	2014/09/02 00:00:06	2014/09/02 00:00:06
$sosrst.xvd	xvd	\	46.1 MB	2014/09/02 00:00:07	2014/09/02 00:01:51	2014/09/02 00:01:51	2014/09/02 00:00:07
$UpCase		\	128 KB	2014/09/02 00:00:06	2014/09/02 00:00:06	2014/09/02 00:00:06	2014/09/02 00:00:06
$Volume		\	0 B	2014/09/02 00:00:06	2014/09/02 00:00:06	2014/09/02 00:00:06	2014/09/02 00:00:06
appswapfile.xvd	xvd	\	2.0 GB	2014/09/02 00:00:15	2014/09/02 00:00:17	2014/09/02 00:00:17	2014/09/02 00:00:15
DeploymentSoftwareDistri...	xvd	\	3.0 GB	2014/09/02 00:00:37	2014/09/02 00:02:00	2014/09/02 00:02:00	2014/09/02 00:00:37
GDVRindex.xvd	xvd	\	101 MB	2014/09/02 00:00:43	2014/09/02 00:02:00	2014/09/02 00:02:00	2014/09/02 00:00:43
ScreenShots.xvd	xvd	\	1.0 GB	2014/09/02 00:00:50	2014/09/02 00:01:21	2014/09/02 00:01:21	2014/09/02 00:00:50
Free space (net)		\	34.7 GB				
Idle space		\					
Volume slack		\	4.0 KB				

Fig. 3. Files under temp content in the stage of restore to factory settings

As shown in Fig. 3, it is found that there is a new format called 'xvd'. According to pervious research [3], files sharing this extension are not the same file type necessarily. Other researchers believed 'xvd' is the abbreviation of Xbox Virtual Disk, which is software package wrapper format utilized by Xbox one to store secured data such as system image or data. It may be used to store console update, system and settings information as well [10].

By comparing timestamps for a new Xbox one and the console restored to factory settings, we find that the created time of the former is May 26, 2013, while the latter is September 2, 2014. It is worth noting that the restore operation was conducted in October 12, 2017. From those information, we can speculate that the operation restored to factory settings gets system files in temp content back to a relatively close version.

Determining the function of ConnectedStorage-retail is another thing we focus on. This file exists in all phases except of restore to factory settings. Based on its name we speculate that this file is related to connect a storage device, such as an inserted flash drive or other external storage. According to timestamps, each operation would lead to the modification of this file. In other words, ConnectedStorage-retail is tightly associated with the change of hard disk.

Another file called appswapfile.xvd attracts our attention as well. We believe it is used to store the contents of swap. There is one area of interest in this file, while it was recreated in the phase of network log on. Therefore we think it is related to memory relocation originated from data writing.

DeploymentSoftwareDistribution.xvd is consider to be utilized for software deployment. In the phase of system update, the modified time of this file went back to September 2, 2014 instead of October 23, 2017, where the operation was conducted. That means DeploymentSoftwareDistribution.xvd concerns system modification only.

3.3 User Content

According to the name, we infer that this partition should be concerned with user information. The contents of root directory under this partition in stage of user log on is demonstrated in Fig. 4.

	Name	Type	Path	Size	Created	Modified	Record changed	Accessed
	[Root directory]				2014/09/02 00:00:07	2014/09/02 00:00:07	2014/09/02 00:00:05	2014/09/02 00:00:05
	SharedStorage		\	0 B	2014/09/02 00:00:05	2014/09/02 00:00:05	2014/09/02 00:00:05	2014/09/02 00:00:05
	PLS		\	0 B	2014/09/02 00:00:05	2014/09/02 00:00:05	2014/09/02 00:00:05	2014/09/02 00:00:05
	$Extend		\	36.0 MB	2014/09/02 00:00:07	2014/09/02 00:00:07	2014/09/02 00:00:07	2014/09/02 00:00:07
	$LogFile		\	64.0 MB	2014/09/02 00:00:07	2014/09/02 00:00:07	2014/09/02 00:00:07	2014/09/02 00:00:07
	$MFT		\	11.4 MB	2014/09/02 00:00:07	2014/09/02 00:00:07	2014/09/02 00:00:07	2014/09/02 00:00:07
	$Bitmap		\	256 KB	2014/09/02 00:00:07	2014/09/02 00:00:07	2014/09/02 00:00:07	2014/09/02 00:00:07
	$UpCase		\	128 KB	2014/09/02 00:00:07	2014/09/02 00:00:07	2014/09/02 00:00:07	2014/09/02 00:00:07
	$Boot		\	8.0 KB	2014/09/02 00:00:07	2014/09/02 00:00:07	2014/09/02 00:00:07	2014/09/02 00:00:07
	$MFTMirr		\	4.0 KB	2014/09/02 00:00:07	2014/09/02 00:00:07	2014/09/02 00:00:07	2014/09/02 00:00:07
	$AttrDef		\	2.5 KB	2014/09/02 00:00:07	2014/09/02 00:00:07	2014/09/02 00:00:07	2014/09/02 00:00:07
	$Secure		\	0 B	2014/09/02 00:00:07	2014/09/02 00:00:07	2014/09/02 00:00:07	2014/09/02 00:00:07
	$Volume		\	0 B	2014/09/02 00:00:07	2014/09/02 00:00:07	2014/09/02 00:00:07	2014/09/02 00:00:07
	$BadClus		\	0 B	2014/09/02 00:00:07	2014/09/02 00:00:07	2014/09/02 00:00:07	2014/09/02 00:00:07
	Free space (net)		\	365 GB				
	Volume slack		\	4.0 KB				
	Idle space		\					

Fig. 4. Files under user content in the stage of user log on

Emphasis would be put on game files concerning this partition. It is shown that four files were created after game installation as shown in Table 2. And the last 3 files are named with strings of hexadecimal digits. Based on the operation conducted, we assume that those files show a close relation to the game itself.

Via viewing files described in Table 2, we find a point that the last 3 files owns the same strings of letters called 'E4EAB7AC-7E08-4571-8BE1-CA99D2C75D45'. According to the record, the game was installed at 10:44 am, August 17, 2018, which is consistent with created time of 3 files discussed above. In addition, the size of Fallout 4 is 26.6 GB, which is quite close to the size of 'E4EAB7AC-7E08-4571-8BE1-CA99D2C75D45'. Therefore, we can speculate that file 'E4EAB7AC-7E08-4571-8BE1-CA99D2C75D45' could be treated as game Fallout 4.

Table 2. Game-related file with its timestamp.

Name	Size	Created time	Modified time
Microsoft.Avatars_8wekyb3d8bbwe.UWA	108 MB	2018/08/17 10:19:22	2018/08/17 10:19:46
e4eab7ac-7e08-4571-8be1-ca99d2c75d45. 22c21d21-1f1b-46b5-916a-a0d690ae61e1	1.6 KB	2018/08/17 10:44:43	2018/08/17 11:15:22
e4eab7ac-7e08-4571-8be1-ca99d2c75d45. de37e6d6-559b-40b0-b1ad-ba2779fbebc4	428 B	2018/08/17 10:44:44	2018/08/17 11:15:22
E4EAB7AC-7E08-4571-8BE1-CA99D2C75D45	27.0 GB	2018/08/17 10:44:43	2018/08/17 11:28:23

Another two files are supposed to be related with game. Further analysis demonstrates that there is no modification for them after playing game. Thus it is concluded that files with the same name as game are related to game initialization.

3.4   System Support

This partition is considered to be related to system as usual. Based on this speculation, tracking changes of files in different system-related stages is critical for the research. It was observed that user.xvd, cms.xvd and WER.xvd changed in various phases as shown in Table 3.

Table 3. Hash Value of specific files.

Name	Wireless network log on	User log on	System update
	Hash value (SHA256)		
user.xvd	F0525A8286B7405C5AA7D CCBE9DBD2E417DD1D5CCA C1CAC936D302568B1B90DC	F422F603C4FA061C9C942 2409545C86CE42868C689 DF7CBC024A3452B8B8AB2A	8FA991CF261B8CF79AFA4E 03E21E7EEAAE198E958174 D585A15DAC9C55250DEC
cms.xvd	C5FEF935144CFE6D5561F 1FA26E0DDF703B2F266EFD 3BE893017F032DBEAE21E	3EC5A59F4EAA36F55A4092 39F69759093D86CF2C6B2 96668225AB1C8369CB86E	/
WER.xvd	F27D3BDA1A1167879461 64107FC4C04E09C589223 1C909236FF111CD8BF0CA4D	261DABFB1DDFF3AF6C22E 512835BEE560EF1D78B8FF 14342040FA23F208E20D6	79622CD942711DC7200934 78F2AF6923767DF93D299 0E240291CC8C05479C028

User.xvd and WER.xvd change in each operation, both system and user operation. As for cms.xvd, previous research [3] has not given a clear definition of it. But it is noted that cms.xvd disappeared since system update. Thus we assume that cms.xvd is no longer utilized in this partition after a certain version.

Similar to those mentioned above, 2 files called E4EAB7AC-7E08-4571-8BE1-CA99D2C75D45.xct and E4EAB7AC-7E08-4571-8BE1-CA99D2C75D45.xvi were created after game installation in this partition. Obviously their names own the same strings as game itself, which means that they were game-related. As we known, xct file type is primarily associated with XVI32, which is a freeware hex editor running under Windows [11]. Besides that, above 2 files have not been changed after playing game, which confirms that they are associated with game initialization only.

3.5   System Update

Even at different stages, this partition is consisted of some fixed files as well. Files in this partition is not that complicated as in others. Details and modifications of files in various phase will be demonstrated in Fig. 5.

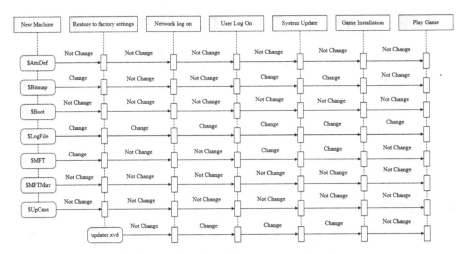

Fig. 5. Modification of files under system update

According to Fig. 5, it is known that updater.xvd is associated with operations conducted to system. Take phase of system update for example, the modified time of this file is changed to October 23 2017, which is corresponding to the latest system update of Xbox one. This explains why there is no change of updater.xvd while Fallout 4 was played.

3.6 System Update 2

This partition seems to be the back-up of System Update. Via comparing hash value of files under root directory, it is found that only $LogFile modified with different manipulations. As we known, $LogFile is a special NTFS system file under Windows. Most research indicate that it is a circular log of all disk operations and used to roll back disk operations. Likewise, $LogFile in this partition is utilized to track disk operations as well.

4 Conclusions

As one popular video game console, Xbox one has attracted numerous game players since its release. Unfortunately, it introduced opportunities and possibilities to illegal and criminal activities as well. Researchers start to show interest in examination and analysis of Xbox one in this situation. Due to its unique operating system, there is no specialized tool for analyzing file or structure in this console. Similarly, few research has been conducted, which leads to few exploration of Xbox one correspondingly.

In this paper we demonstrate a new way to explore Xbox one. Via performing a series of operations and comparing change towards the same file, rules and features were summarized so as to give a deep understanding of the console. Then further analysis was conducted, including partition lay out, file system, data format, system file as well as specific files under root directory. And it is envisioned that more research about new Xbox series is likely to be conducted in the future.

Acknowledgments. This paper is supported by National key research and development plan. The People's Republic of China ministry of science and technology, project number: 2017YFC0803805.

References

1. Information on https://en.wikipedia.org/wiki/Xbox_One
2. Information on https://wccftech.com/xbox-one-39-1m-units-estimated/
3. Moore, J., Baggili, I., Marrington, A., Rodrigues, A.: Preliminary forensics analysis of the Xbox One. Digital Invest. **11**, 57–65 (2014)
4. Podhradsky, A.L., D'Ovidio, R., Casey, C.: A practitioners guide to the forensic investigation of Xbox 360 gaming consoles. In: ADFSL Conference on Digital Forensics, Security and Law (2011)
5. Bolt, S.: XBOX 360 Forensics: A Digital Forensics Guide to Examining Artifacts (2011). (ISBN 978-1-59749-623-0)
6. Shannon, M.M.: Forensic relative strength scoring: ASCII and entropy scoring. Int. J. Digit. Evid. **2**(4), 151–169 (2004)
7. Information on https://www.logicube.com/shop/falcon/?v=1c2903397d88
8. Information on http://www.x-ways.net/forensics/index-m.html
9. Information on https://karens-hasher.en.lo4d.com/windows
10. Information on https://fileinfo.com/extension/xvd
11. Information on https://fileinfo.com/extension/xct

Online Learning Engagement Assessment Based on Multimodal Behavioral Data

Liying Wang[1(✉)] and Yunfan He[2]

[1] School of Education Science, Nanjing Normal University, Nanjing 210097, China
wangliying@njnu.edu.cn
[2] Graduate School of Education, Peking University, Beijing 100871, China
heyunfan@pku.edu.cn

Abstract. Online learning monitoring is still on the urgent demand to track and analyze the learning engagement of the learners. To this end, the multimodal behavioral data of online learners are collected during the online learning process from the psychological, physiological and behavioral dimension which are respectively dependent on the techniques of expression recognition, physiological heart rate acquisition with Internet of Things and operation events listening. Based on the dataset with 25 experimental p, online learning engagement states are statistically analyzed and assessed by applying the schema including both prior rules and data fitting method which can quantitatively evaluate the learning engagement. Through the questionnaire survey and statistical validation, the assessment results show the schema can measure learning engagement in multiple dimensions and evaluate the whole state more comprehensively and automatically.

Keywords: Online learning · Learning engagement · Multimodal behavioral data · Facial expression recognition · Internet of Things technology

1 Introduction

With the rapid development of MOOCs world widely, online learning is becoming more and more popular and important in higher education and corporate training. At the same time, some research have revealed that specific problems still exist such as less interaction with companions and teachers in the learning process, more confusion about the learning contents, high dropout rate and low utilization rate.

The main reason for these problems lies in the fact that online learning dramatically depends on individual autonomy. The cognitive, emotional and behavioral changes of the learners in the whole learning process are still invisible to a great extent for teachers. At present there lacks deep and continuous tracking of online learning engagement, such as meaningful behaviors assessment, especially necessary emotional state description and mental activity analysis. Therefore, it is difficult to diagnose the learning progress and predict the academic achievement automatically.

In the recent twenty years, the studies on student engagement integrating with informatics and sensor techniques to measure the engagement level have made significant progress in education field. The contributions of this paper have three points. Firstly,

© Springer-Verlag GmbH Germany, part of Springer Nature 2020
Z. Pan et al. (Eds.): Transactions on Edutainment XVI, LNCS 11782, pp. 256–265, 2020.
https://doi.org/10.1007/978-3-662-61510-2_25

a multimodal data acquisition architecture is designed to collect multi-learner's multi-channel data in parallel in online learning process. Secondly, according to multimodal data, semi-automatic tracking and quantitative analysis are executed for every learner every channel. Thirdly, an assessment schema, including both prior rules and data fitting method, is proposed. It is based on the three primary learning theories of behaviorism, cognitivism and constructivism, the explicit physiological responses, psychological responses and behavioral responses can be the basic cues to understand the learning engagement state of the learners. Through online learning engagement assessment, it is helpful for teachers and learners to adjust learning strategies and improve learning effects as far as possible.

The rest of the paper is organized as following. Section 2 explains the related concept of online learning engagement and learning analytics techniques. Section 3 introduces our multimodal behavior dataset collection schema. Section 4 describes the assessment schema for online learning engagement in detail. Finally, the experimental results are exhibited in Sect. 5.

2   Online Learning Engagement and Assessment

To assess the learning engagement of online learners reasonably, the related concept of online learning engagement and the measurement techniques of learning analytics are explained thoroughly in this section.

2.1   Online Learning Engagement

From the engagement as the beginning, Schaufeli [1] defined engagement as a positive, fulfilling, and work-related state of mind that is characterized by vigor, dedication, and absorption.

School engagement has the multifaceted nature as Fredricks [2] pointed out, which can be split into three dimensions of behavioral engagement, emotional engagement and cognitive engagement.

Learning engagement usually means the emotional state, cognitive participation state and behaviors of students generated by interacting with the learning environment during the learning process [3].

In the past, the research on online learning engagement mainly applied the self-report, survey methods or single-modal sensor to collect data, conduct statistics, and analyzed its component dimensions and influencing factors.

In 2011, on the basis of a consensus to characterize student engagement during learning activities as a three component construct featuring behavioral, emotional, and cognitive aspects, Reeve [4] proposed agentic engagement as an essential new aspect which was defined as students' constructive contribution into the flow of the instruction they receive. In 2012, Dixson [5] discovered multiple communication channels may be related to higher engagement and that student-student and instructor-student communication are strongly correlated with higher student engagement with the course.

In Kahu [6] focused on two key elements of emotional engagement: interest and belonging. Findings highlight the importance of interest triggered by personal preferences and experiences. Interest led to enjoyment, increased behavioral engagement with greater time and effort expended, and improved cognitive engagement in terms of depth and breadth of learning. Compared to face to face courses and online forums, distance study has more need for staff to consider emotional engagement when designing and delivering the curriculum and when interacting with students to reduce the sense of isolation. In Lee [7] discussed autonomy support as a strategy to enhance online students' intrinsic motivation and engagement, he proposed three guidelines to provide choices, rationale behind why assignments are designed in particular ways, and flexibility in completing more personally meaningful assignments, so as to engage students in active participation and successful completion of the course.

These studies show that the external environment has an important impact on the online learning engagement, and it is necessary to set up the environment to impact on the behavior of the learners and improve the learning effect.

As Christenson et al. [8] argued although researchers have been reached consensus that student engagement is multidimensional, multidimensionality differs from the number and types of engagement dimensions. Learning engagement is comprehensive, can be understood as the learning outcome, a process to other desired outcomes, or plays a dual role. Researchers will need to define clearly own conceptualization in each study according to the definite real learning environment and specific teaching strategy.

2.2 Technologies for Learning Analytics

Learning analytics (LA) is sprouting since 2010. Siemens [9] defines LA is the technique which applies data mining and analysis model to discover the internal information and relationship, to predict and improve learning. The use of analytics in education has grown in recent years for four primary reasons: a substantial increase in data quantity, uniform data formats, advances in computing, and increased sophistication of tools available for analytics. With the development of hardware and software of the Internet of things increasingly convenient and cheap, the means of collecting physiological and psychological data for cognitive and emotional tracking are feasible and widespread.

From 2016, the deep neural networks learning algorithms are gradually becoming matured. It has been applied in more and more Artificial Intelligence (AI) field. Nowadays, because the accuracy of face detection and face recognition has reached over 90%, many applications such as electric payment and security check can depend on human face identification. At the same time, well-known companies have provided open API to get access to their technologies, such as Google face service, Microsoft cognitive services, Baidu brain, Face++ [10] and so on. In addition, open-source algorithms are easily accessible to develop for own applications, such as Openface, Fast RCNN on GitHub, and so on. At present, AI technologies affect education field deeper and deeper.

Zhang et al. [11] discussed the learning engagement has the dynamic, multidimensional and contextual characteristics. Therefore, in a real learning situation, multimodal data means comprehensive multiple data types with rich learning patterns. The analysis of multiple-modality data, combined with machine learning method, both the fine-grained indicators of learning engagement and the modeling process in different scenarios can be analyzed. From the literature review, the conclusion can be drawn that multimodal sensors are better than single sensor in the aspect of accuracy for learning analytics.

3 Multimodal Behavior Data Perception

As the basis of the online learning engagement assessment, it is necessary to automatically perceive the multi-dimensional learning data in an online learning space-time scenario. In order to track the three dimensions of cognitive, emotional and behavioral state, The corresponding multimodal data of the learners in online learning process include the data of facial video, heart rate, mouse operation, keyboard operation, process operation and so on.

As shown in Fig. 1, the multimodal behavior data flow chart illustrates the process of online learning perception. The collected data and further statistical data about the learning engagement state are all integrated and stored into the dataset.

3.1 Data Collection Schema

The collection schema integrates the hardware of the camera and heart rate sensor, and develops the software components of video capture, heart rate collect, listeners of mouse keyboard and process operation. The multimodal data are saved into the corresponding data format for later analysis.

The video capture tries different hardware devices to capture videos, such as webcam and desktop camera. The video sequences are stored in the MP4 format.

Heart rate collector is a set of real-time monitoring heart rate components using the Internet of Things (IoT) technology. The front end of the collector is made up of Arduino NodeMcu ESP8266 development board and Pulsesensor heart rate sensor. At the back end Sitewhere is used as the Internet of things server and MongoDB as a data storage server. The Pulsesensor should be fixed on one finger surface of the learner, and its power supply line should be connected to the USB slot. As long as the pulsesensor is powered and attached to the user's finger, the heart rate data can be sent to sitewhere server and saved into MongoDB.

The operation behavior collector includes mouse, keyboard and process listener. Mouse listener records mouse click and release operations in the learning process. Keyboard listener records the contents of keyboard input. The process listener records the process information of the operating system once every 15 s: including the process name, the state, the create time, the thread number, etc. These data are appended to their respective text logs in the format of time and content.

3.2 Data Statistics Schema

The original data pass different process path according to the three aspects of cognition, emotion and behavior, so as to assess the learning engagement of the learner.

Facial expression recognition is used to get the emotion state of the learner. Firstly, the learner's complete online learning video is transformed from MP4 into JPG images, using the FFmpeg open source tool to extract a key frame every five seconds. Secondly, the open API of Face++ facial expression recognition is called with one image input and one expression label output.

The statistical data of heart rates are used to get the cognitive state of the learner. The first step is to get heart rate data from the Sitewhere server via the HTTP protocol in

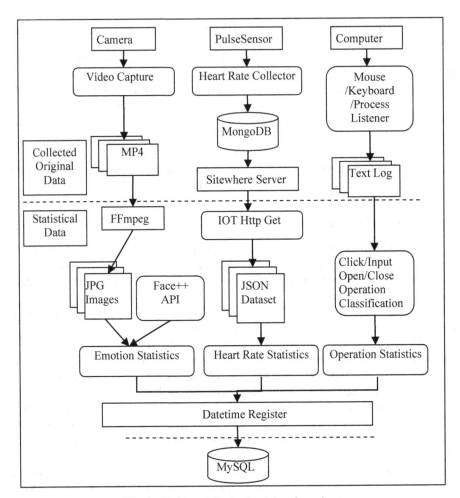

Fig. 1. Multimodal behavioral data flow chart

JSON data format. The second step is to calculate the mean value per second according to the time series.

The statistical data of computer operations come from the text logs, through counting the click times of the mouse, the input content of the keyboard, the process name of the opening and closing between two adjacent moments. Finally, it counts the number of keyboard, mouse and the number of opening and closing processes.

All the statistics data are stored into the MySQL database server.

3.3 Online Learning Experiment Schema

The experimental subjects come from students of educational technology major taking the course of "network security and maintenance". Before the test, the data collection agreements had been signed with the subjects to protect the privacy of the students.

The online course "computer network" produced by Harbin Institute of Technology on the Chinese University Moocs platform [12] is adopted as the experimental inducement materials. The experiment collects behavioral data of 36 participants in online learning a particular chapter.

After the learning, the survey on learning behavior and engagement of learners is done using one questionnaire. After data integrity and effectiveness analysis, the data of 25 participants are valid and assessed in the following steps.

4   Online Learning Engagement Assessment

Through tracking and analyzing personal data, learners' self-diagnosis and overall report of learning engagement are assessed. Two ways used to assess their engagement are prior rules and data fitting method.

4.1   Prior Assessment Rules

Three dimensions of emotional, cognitive and behavioral engagement are assessed separately and then combined together to assess the learning engagement. As shown in Table 1, we define prior rules to judge the learners' engagement state, in which the threshold of the judgments as the parameters need be tuned by users.

Firstly, there are four types of learning emotion: positive, negative, distracted and mixed. Eight basic emotion types can be obtained by the face++ expression recognition tool which analyzes key frame images of the facial video and marks the emotion types.

Table 1. Learning engagement assessment prior rules

Engagement dimension	Types	Rules
Learning Emotion	Active	The sum of happiness, surprise and neutral ratio is greater than or equal to 75%
	Passive	The sum of sadness, anger, disgust and fear is greater than or equal to 50%
	Distracted	No Face ratio is greater than or equal to 40%
	Mixed	Else
Learning Psychology	Stable	The proportion of the Stable interval is equal to 70%
	Fluctuated	Else
Learning Behavior	Good	The ratio of operation related to learning is more than 70% and keyboard input is more than twenty
	Warning	The ratio of operation related to learning is greater than 50% and less than 70%, and the keyboard input is less than twenty
	Bad	The ratio of operation related to learning is less than 50% or no operation

Thus the sequence of emotion labels is obtained and grouped into four types of learning emotion.

Secondly, as the indicator of cognitive engagement, the ratio of stable heart rate was calculated. Taking the average heart rate as the reference value R, the interval [0.9R, 1.1R] is labeled as stable psychology, the range of which could be defined more accurately according to related psychological theories. The heart rate is higher than the upper bound or lower than the bottom bound is labeled as fluctuated one.

Finally, the ratio of operation related to learning and the number of the keyboard is counted, the behavior types include good, warning and bad label.

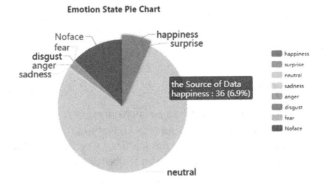

Fig. 2. Emotion state pie chart

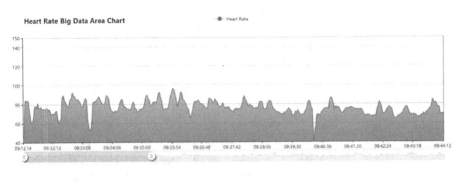

Fig. 3. Psychology state area chart

Here, a participant in the online learning experiment is taken as an example. His learning engagement assessment results are shown in below figures below. Figure 2 shows the ratio of different learning emotion types in the pie chart. As the active emotion ratio is more than 75%, the evaluation result is good.

Figure 3 shows the heart rate distribution. Figure 4 shows the number of operations per minute. According to the questionnaire and talk survey, this learner felt engaged in this online learning experiment. This is coincident with the good result of the emotion state assessment. But he didn't concentrate on learning enough because he did many operations unrelated to learning and didn't input any characters related to learning.

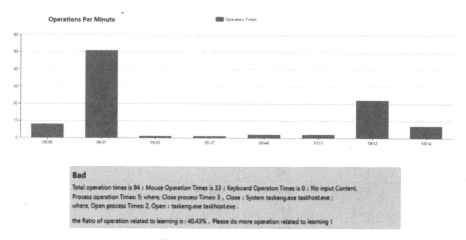

Fig. 4. Behavoir state bar chart

4.2 Learning Engagement Data Fitting

In order to predict the score of the learning engagement, the supervised machine learning methods can be adopted by training on the samples dataset. Supposed the learning engagement score $E = \sum wx$, where x is the score vector with three dimensions of emotional, cognitive and behavioral engagement, which are calculated from statistical data,E is the score from the questionnaire survey data. Data fitting by the least square method is done to get the weight vector. Based on our dataset with 25 experimental participants, the weight vector w is (0.017161, 0.30504, −0.00035).

All statistical data and survey data are shown in Table 2, in which the duration is the online learning time of each participant. Finally the average variance error between the questionnaire and predicted scores of the learning engagement is 0.9667, which is less than 1.

Table 2. The statistical and survey data of learning engagement

ID	Duration (min:sec)	Engagement score (E)	Facial image number (x1)	Heart rate average (x2)	Operation count (x3)	Assessed engagement
1	32:20	29	383	83.48	198	31.97
2	43:40	24	300	82.33	208	30.19
3	55:30	37	556	87.46	349	36.10
4	47:10	34	440	85.5	66	33.61
5	39:50	34	395	89.56	3920	32.72
6	43:25	34	454	71.34	38	29.54
7	50:55	27	518	77.44	197	32.44
8	50:00	41	553	81.5	106	34.31
9	49:45	40	411	89.92	449	34.32
10	48:35	32	558	83.09	132	34.88
11	46:35	30	393	88.24	154	33.61
12	56:10	34	251	69.62	606	25.33
13	40:50	32	383	65.59	66	26.56
14	48:45	24	347	76.38	303	29.15
15	40:05	33	268	66.68	181	24.88
16	53:20	25	531	77.1	0	32.63
17	59:10	30	458	87.06	198	34.35
18	47:40	32	344	71.58	267	27.64
19	37:15	25	273	83.15	151	30.00
20	46:55	25	344	71.82	119	27.77
21	39:45	32	345	74.01	428	28.35
22	43:00	29	263	94.25	427	33.11
23	44:00	32	305	70.86	144 ˙	26.80
24	42:45	30	264	79.01	256	28.54
25	38:20	31	359	96.01	9663	32.05

5 Conclusion

The multimodal online learning behavior analysis is beneficial to assess the learning engagement in quantity. Our experimental schema is verified in the actual online learning process. The questionnaire survey and statistical validation show that it has low invasion and few interferences to online learners, and can fully perceive and assess learning engagement. The experimental results prove the assessment schema can measure learning engagement in multiple dimensions and evaluate the whole state more comprehensively and automatically.

Albeit so, there are still some limitations about our method which could be improved in the future work. One is a more comprehensive structure of online learning engagement should be interpreted and constructed based on the educational meaning. The other is more machine learning predictive models should be investigated and compared.

From these data, we find out that the online course in video format is helpful and easy for learners to acquire professional contents. But the learner still lies in the passive acceptance situation. The MOOCs lacks face to face interactive discussion to activate deep critical thinking, not to mention to construct personal own and community common knowledge. In the future work, it is possible for us to make use of the perception schema, give the learners proper interaction at the inattentive or confused occasion.

Acknowledgments. The paper is supported by the education science plan foundation "in 12th Five-Year" of Jiangsu province of China (B-a/2015/01/010).

References

1. Schaufeli, W.B., Martinez, I.M., Pinto, A.M., Salanova, M., Bakker, A.B.: Burnout and engagement in university students: a cross-national study. J. Cross Cult. Psychol. **5**, 464–481 (2002)
2. Fredricks, J.A., Blumenfeld, P.C., Paris, A.H.: School engagement: potential of the concept, state of the evidence. Rev. Educ. Res. **1**, 59–109 (2004)
3. Zhang, Q., Wang, H.: Multiple-modality data representation of learning engagement: supporting theory, research framework and key technologies. e-Educ. Res. **12**, 1–8 (2019). (in Chinese)
4. Reeve, J., Tseng, C.-M.: Agency as a fourth aspect of students' engagement during learning activities. Contemp. Educ. Psychol. **4**, 257–267 (2011)
5. Dixson, M.D.: Creating effective student engagement in online courses: what do students find engaging? J. Scholarsh. Teach. Learn. **2**, 1–13 (2010)
6. Kahu, E.: Increasing the emotional engagement of first year mature-aged distance students: interest and belonging. Int. J. First Year High. Educ. **2**, 45–55 (2014)
7. Lee, E., Pate, J.A., Cozart, D.: Autonomy support for online students. TechTrends **4**, 54–61 (2015)
8. Christenson, S.L., Reschly, A.L., Wylie, C.: Handbook of Research on Student Engagement. Springer, New York (2012)
9. Sawyer, R.K.: Cambridge Handbook of the Learning Sciences, 2nd edn. Cambridge University Press, Cambridge (2014)
10. https://www.faceplusplus.com.cn/
11. Zhang, Q., Wu, F., Xu, W.: Learning engagement evaluation supported by multimodal data: status, implications and research trends. J. Distance Educ. **1**, 76–86 (2020). (in Chinese)
12. https://www.icourse163.org/course/HIT-154005

Distributed Cache and Recovery Method for Strong Real-Time Applications

Qing Cai[✉], Jiabo Lu, and Mingda Lei

Jiangsu Automation Research Institute, Lianyungang, Jiangsu, China
bearangel2001@163.com
http://www.716.com.cn/Portal/Index.aspx

Abstract. Failures (including hardware failures, software failures or unexpected shutdowns, sudden power failures, etc.) are unavoidable problems in large-scale distributed systems, so current distributed systems are required to support systematic fault tolerance. Aiming at the requirements of strong real-time application scenarios, this paper proposes a distributed cache and recovery method based on memory database. SQLite memory database is adopted and election-based multi-node data synchronization is introduced, which ensures the strong consistency of data on each node and eliminates the bottleneck and failure problems caused by the setting of the central node; at the same time, a dynamic load balancing mechanism is adopted to reduce the amount of synchronized data in the entire system and ensure the smooth operation of the system. Finally, the effectiveness of the proposed method is proved by experiments.

Keywords: Distributed systems · Strong real-time applications · SQLite memory databases · Load balancing

1 Overview

Failures (including hardware failures, software failures or unexpected shutdowns, sudden power failures, etc.) is an inevitable problem in large-scale distributed systems. At present, the fault tolerance of the system is emphasized in the design process of the current distribution system. When some hardware (software) of a distributed system fails, a fault tolerance mechanism can be used to prevent a large impact on the system [3]. For strong real-time applications, the fault tolerance of the system [4] is more important, such as strong real-time control for naval gun weapons, ship-to-air missile weapons and so on [14–16]. There are strict time requirements for application processing delay, data consistency, data transmission sequence and system response time. A short system pause will also lead to control failure and affect the use of the whole system. Therefore, its very urgent to recover from the fault for strong real-time applications [17–19].

In distributed systems, there are many fault recovery strategies to improve the fault tolerance of the system. For example, the frequently used dual-machine

© Springer-Verlag GmbH Germany, part of Springer Nature 2020
Z. Pan et al. (Eds.): Transactions on Edutainment XVI, LNCS 11782, pp. 266–274, 2020.
https://doi.org/10.1007/978-3-662-61510-2_26

hot backup strategy [5–7], for critical applications that require key guarantees, the same critical application is run on two nodes at the same time, one is the primary node and the other is the secondary node. In the event of a failure of any node, the other node can ensure that the system can continue to run [8]. However, the shortcomings of this strategy are as follows: firstly, dual backup nodes need to be set in advance, which is not flexible enough; secondly, dual computers run the same application at the same time, which is a waste of computing resources. It is difficult to provide sufficient resources for the compact system; thirdly, there is still a certain possibility that both computers may fail, and other nodes cant be used for fault replacement. There are many other fault recovery strategies [9–11], but still fail in dealing with the requirements of strong real-time applications.

In this paper, aiming at the fault-tolerant requirements of strong real-time applications, a distributed cache and recovery method based on memory database is proposed. Firstly, it is based on SQLite [12,13] memory database to improve the efficiency of data reading and writing. Secondly the method realizes automatic synchronization of multi-node data based on election, which ensures the strong consistency of data on each node and eliminates the bottleneck and failure problem caused by the setting of a central node. Thirdly, the use of dynamic load balancing mechanism [20,21] which can not only reduce the amount of synchronous data of the whole system but also ensure the smooth operation of the system [22].

2   Model Structure

The distributed cache and recovery method for strong real-time applications is divided into three levels from top to bottom, the load balancing layer, the synchronous recovery layer, and the node data layer.

2.1   Implementation Method of Node Data Layer

The data node uses an SQLite database. SQLite is an open-source database engine. It has the advantages of a small amount of code, small memory, no configuration, support SQL, has ACID characteristics, good portability. It can be well adapted to scenarios with frequent read operations, less write operations, and data synchronization between the client and server. The purpose is to improve the efficiency of database query operations as much as possible.

2.2   Implementation Method of Synchronous Recovery Layer

The synchronous recovery layer is mainly aimed at the data synchronization of multiple nodes, and it is achieved by active push between non-center memory databases. The main steps are in Fig. 1.

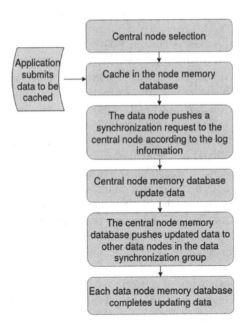

Fig. 1. Main steps of the implementation method of synchronous recovery layer.

Selection of Central Node. There is a central node in the data synchronization group, and all data updates are aggregated to the central node for combination and distribution. The selection of the central node is mainly produced by competition [23, 24], and the comprehensive evaluation is performed according to the node capabilities, namely the average CPU load, the average memory load, and the average network load. The node gained the highest scores assumes the central node. Specifically, let $W_{CPU}, W_{Mem}, W_{Net}$ denotes the weight coefficient of C(average CPU load), M(average memory load), N(the average network load) separately. The storage cost C_i of data node DN_i can be defined as

$$C_i = W_{\mathrm{CPU}} \times C + W_{\mathrm{Mem}} \times M + W_{\mathrm{Net}} \times N. \tag{1}$$

When the data node starts, it automatically starts the central node service. If not, activate local central node service and set it as the central node service. Each data node samples information every t seconds, including the usage of CPU run time, CPU idle time, memory amount, memory usage, network upstream traffic, and network downstream traffic. Calculate the value C_i of this node, and sends it to the central node. If there is already a central node when a new node joins the data synchronization group, the current central node is used. If the separation and combination of data synchronization groups result in multiple central nodes, choose the central node with highest C_i score as the current central node.

Synchronization Strategy. The updates from nodes to data are first passed to the central node and then propagated to other data nodes.

(a) After the central node receives and confirms data update, it will notify the data nodes to update;

(b) The data nodes periodically communicate with the central node to query the data status;

(c) If a conflict exists when nodes update data, update data nodes immediately.

Strong Consistency Guarantee Strategy. All data updates go through the central node which can guarantee the strong consistency of the data.

(a) At any time, the data status of the central node is the final status of the data;

(b) When the data of the data node lags behind the central node, the data can still be updated. but:

1. Data node cannot add entries that already exist on the central node. The entries are determined by the primary key, and the same primary key is considered as the same entry;

2. Data node cannot delete entries that do not exist on the central node;

3. Data node cannot modify entries that have been modified on the central node, that is when the entries of the central node and the entries of the data node are inconsistent, the central node cannot be modified.

(c) When data node lags and causes an update conflict, the data will be rolled back instead of being submitted to the data node's database;

(d) When data node lags and causes update conflicts, they actively synchronize with the central node.

2.3   Implementation Method of Load-Balanced Layer

The load-balanced layer adopts the architecture of nginx + keepalived + LVS + DNS, which can well solve the problems of high availability, scalability, reverse proxy and extension balancing. The architecture is shown in Fig. 2.

LVS/nginx [1] uses cluster technology to get a high performance, high availability, load balancing server at the Linux operating system level. Multiple nginx and tomcat can be extended with LVS/nginx (The LVS proposed above can be replaced with nginx).

Deploy two LVSs to form a cluster, and deploy keepalived on each LVS. Then set up to the same virtual IP to ensure the high availability of LVS. When one LVS fails, keepalived [2] can detect it and automatically migrate traffic to another LVS. This process is transparent to the user. Nginx is internally controllable and can easily expand its capacity at any time by increasing the number of web-servers. This ensures the availability of the site layer, and nginx can migrate traffic to other tomcats if any tomcat fails. DNS polling allows the LVS layer to expand horizontally, enabling performance expansion by adding machines.

Characteristics of load-balanced architecture are listed below:

(a) expand the performance of LVS entrance layer linearly through DNS polling

(b) ensure high availability by keepalived

(c) extend multiple nginx through LVS

(d) balance the load through nginx, and conduct seven-layer routing

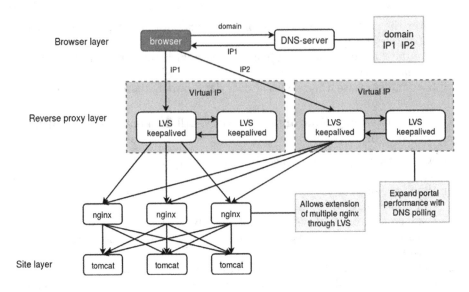

Fig. 2. The architecture of the load-balanced layer.

3   Simulation Verification

In order to verify the effectiveness of the proposed method, the following two sets of performance testing experiments are carried out.

Performance test 1: According to the performance test requirements, we set up 6 servers on Linux and VxWorks virtual machines respectively. After running the test program locally, we can get the time and speed of the database create operation, update operation, and delete operation under the Linux environment as shown in Table 1. The corresponding test speed on the VxWorks virtual machine is shown in Table 2.

Table 1. The speed of the database create operation, update operation, and delete operation under the Linux environment.

	Create	Delete	Update
50	2	5	2
100	4	6	4
500	11	20	17
1000	24	34	36
5000	113	150	126
10000	226	316	343
50000	1171	1852	1888

From the performance test experiment 1, it can be found that our scheme is better than the comparative scheme.

Table 2. The speed of database create operation, update operation, and delete operation under VxWorks environment.

	Create	Delete	Update
50	17	17	66
100	17	33	0
500	33	34	16
1000	50	67	50
5000	233	476	250
10000	430	534	500
50000	2933	2666	2563

Table 3. The Windows virtual machine operating time. (10 node connections)

	Receiving time	Process time	Communication time
0	0	0	
1000	0	0	
2000	359	78	281
3000	625	203	63
4000	890	141	124
5000	1156	172	94
6000	1406	219	31
7000	1873	219	248
8000	2484	516	95
9000	2953	328	141
10000	3453	500	0

Performance Test 2: According to the performance test requirements, we set up 10 servers and 10 corresponding clients on windows, Linux and VxWorks virtual machine respectively. We create 1000 items every client and test the speed of this create operation. During the experiment, 10 windows virtual machines are successfully connected and operated. The synchronization speed is shown in Table 3. At the same time, 10 Linux virtual machines are successfully connected and operated. The synchronization speed is shown in Table 4. However, when 10 VxWorks virtual machine nodes are connected, only four clients can be started simultaneously at the most. Otherwise, the rc = dist_open_memdb (argv[1], &db) function will get value 1, which will make the program exited. So

Table 4. The Linux virtual machine operating time. (10 node connections)

	Receiving time	Process time	Communication time
0	0	0	
1000	0	0	
2000	91	33	58
3000	885	27	767
4000	1016	35	96
5000	1777	24	737
6000	2667	24	866
7000	3143	27	449
8000	3548	26	379
9000	3683	23	112

Table 5. VxWorks virtual machine operation time. (4 nodes)

	Receiving time	Process time	Communication time
0	0	0	
1000	0	0	
2000	100	34	66
3000	150	33	17
4000	216	34	32
5000	266	33	17
6000	316	33	17
7000	383	34	33
8000	466	33	50

In the experiment, only four service nodes are tested. We test the speed of 8000 create operations, as shown in Table 5.

4   Conclusion

For fault-tolerant recovery after large-scale distributed system fault (including hardware failure, software failure, unexpected shutdown, sudden power failure, etc.), we proposed a distributed cache and memory database recovery method based on memory database. Implemented by SQLite memory database, our method introduces automatic election-based multi-node data synchronization to ensure strong consistency of data on each node and meanwhile eliminates bottlenecks and failures caused by the setting of the central node. At the same time, a dynamic load balancing mechanism can reduce the synchronization data of the whole system and ensure the smooth operation of the system. Finally, the effectiveness of the proposed method is proved by our experiments.

References

1. Nginx. http://nginx.org
2. Keepalived. https://www.keepalived.org
3. Ramakrishnan, R., et al.: Azure data lake store: a hyperscale distributed file service for big data analytics. In: The 2017 ACM International Conference. ACM (2017)
4. Liao, C., Squicciarini, A., Lin, D.: LAST-HDFS: location-aware storage technique for hadoop distributed file system. In: IEEE International Conference on Cloud Computing. IEEE (2017)
5. Kakoulli, E., Herodotou, H.: OctopusFS: a distributed file system with tiered storage management. In: ACM International Conference. ACM (2017)
6. Choi, W.G., Park, S.: A write-friendly approach to manage namespace of Hadoop distributed file system by utilizing nonvolatile memory. J. Supercomput. **75**(10), 6632–6662 (2019). https://doi.org/10.1007/s11227-019-02876-9
7. Busca, J.-M., Picconi, F., Sens, P.: Pastis: a highly-scalable multi-user peer-to-peer file system. In: Cunha, J.C., Medeiros, P.D. (eds.) Euro-Par 2005. LNCS, vol. 3648, pp. 1173–1182. Springer, Heidelberg (2005). https://doi.org/10.1007/11549468_128
8. Soules, V., et al.: Distributed cache management in information-centric networks. IEEE Trans. Netw. Serv. Manage. **10**(3), 286–299 (2013)
9. Cardenas, Y., Pierson, J.M., Brunie, L.: Uniform distributed cache service for grid computing. In: Sixteenth International Workshop on Database and Expert Systems Applications 2005, Proceedings. IEEE (2005)
10. Zhang, J., Li, Q., Zhou, W.: HDCache: a distributed cache system for real-time cloud services. J. Grid Comput. **14**(3), 407–428 (2016). https://doi.org/10.1007/s10723-015-9360-9
11. Gao, X., Fang, X.: High-performance distributed cache architecture based on Redis. Lecture Notes in Electrical Engineering, vol. 270, pp. 105–111 (2014)
12. Junyan, L., Shiguo, X., Yijie, L.: Application research of embedded database SQLite. In: International Forum on Information Technology and Applications 2009, IFITA 2009. IEEE (2009)
13. Owens, M.: Embedding an SQL database with SQLite. Linux J. **2003**(110), 2 (2003)
14. Shi-Yan, S., Zhi-Ming, Q.: A comparison study on optimal configuration methods of naval gun weapon systems. Acta Armamentarii **28**(7), 778–781 (2007)
15. Sun, S., et al.: A study on the optimal design method of naval gun weapon system. In: Control & Decision Conference. IEEE (2008)
16. Liu, X., et al.: Fault diagnosis for hydraulic system of naval gun based on BP-Adaboost model. In: 2017 Second International Conference on Reliability Systems Engineering (ICRSE). IEEE (2017)
17. Xu, J., et al.: Approach for combat capability requirement generation and evaluation of new naval gun. In: 2017 36th Chinese Control Conference (CCC) (2017)
18. Guoqiang, L., et al.: Study on a fire distribution model of integrated naval gun and laser weapon system. In: The 30th Chinese Control and Decision Conference (2018)
19. Huang, Y., et al.: The study on the optimal filtering length for closed-loop spotting of close-in anti-missile naval gun weapon system. In: International Conference on Computer Application & System Modeling. IEEE (2010)
20. Stanković, R., Štula, M., Maras, J.: Evaluating fault tolerance approaches in multi-agent systems. Auton. Agents Multi-Agent Syst. **31**(1), 151–177 (2015). https://doi.org/10.1007/s10458-015-9320-6

21. Arabnejad, H., Pahl, C., Estrada, G., Samir, A., Fowley, F.: A fuzzy load balancer for adaptive fault tolerance management in cloud platforms. In: De Paoli, F., Schulte, S., Broch Johnsen, E. (eds.) ESOCC 2017. LNCS, vol. 10465, pp. 109–124. Springer, Cham (2017). https://doi.org/10.1007/978-3-319-67262-5_9

22. Fischer, M.J., Lynch, N.A., Paterson, M.S.: Impossibility of distributed consensus with one faulty process. J. ACM **32**(2), 374–382 (1985)

23. Lamport, L.: Fast Paxos. Distrib. Comput. **19**(2), 79–103 (2006). https://doi.org/10.1007/s00446-006-0005-x

24. Abraham, I., et al.: Byzantine disk paxos: optimal resilience with byzantine shared memory. Distrib. Comput. **18**(5), 387–408 (2006). https://doi.org/10.1007/s00446-005-0151-6

Cross Cultural Hierarchy Phenomenon: A New Communication Mechanism to Disseminate Chinese Culture Overseas Based on Social Media

Jun Shen[✉]

School of Journalism and Communication, Nanjing Normal University, Nanjing 210024, China
Ekky74@126.com

Abstract. Culture is created during the process of human activities, and the development and prosperity of culture highly depend on the dissemination of media. In the context of international communication, Chinese culture is facing up with challenges both from cultural prejudice and media bias. With the breakthrough of Web 2.0, new media technologies have subverted the communication mechanism of the traditional media, and cultural communication also changed dramatically. Since the Strategy of Chinese Culture Going-out is launched, it is high time to adopt new technology and creative thinking to promote Chinese culture overseas. This paper explores the communication mechanism of social media, by analyzing the three core components of Platform, User and Content of social media. Based on the principle of Acculturation, it posits that the key points to break personal information cocoon should include the following methods: 1. Enriching cultural communication forms (utilizing the characteristic of social platform); 2. Customizing cultural communication scheme (by User Typology analysis); 3. Creating cultural communication contents (converging multiple new media technologies). This paper further points out that a Cross Cultural Hierarchy Phenomenon would occur if before-mentioned methods are adopted to raise cultural communication effect, and it is believed that the passive situation of Chinese cultural communication should be changed to promote Chinese culture overseas.

Keywords: Social media · Chinese cultural communication · Cross Cultural Hierarchy Phenomenon

1 Introduction

Since China launched the Policy of Culture Go-out, Chinese government has implemented an active cultural diplomacy, whilst it didn't change the stereotype of Chinese traditional culture, and even seemed to cause prejudice to some extent. As an example, Chinese government extensively set up CIs (Confucius Institutes) which co-operated with universities all over the world, with the aim of stepping up cultural exchanges with the rest of the world in a joint promotion of cultural prosperity, but it was blamed to

J. Shen—This paper is supported by "Research on the Concept of Community of Shared Future of Mankind and the Reconstruction of Global Communication Order" (18BXW062).

© Springer-Verlag GmbH Germany, part of Springer Nature 2020
Z. Pan et al. (Eds.): Transactions on Edutainment XVI, LNCS 11782, pp. 275–286, 2020.
https://doi.org/10.1007/978-3-662-61510-2_27

"interfere with academic freedom". Even though Chinese mainstream media such as Xinhua News Agency quoted president Xi jinping's proposition on CIs, that is, "Confucius Institutes belong to China and the world as well," while still the criticism was stated as "China's government is gaining a foothold on American campuses" [1].

One factor that cause the prejudice lies in different culture identities and cultural beliefs. Xi's proclaim stems from traditional culture belief "harmony but not uniform" [2], constructing a harmonious status of integrating different cultures, which can hardly be approved by people who hold the belief of Clash of Civilization. Apart from different culture identities and cultural beliefs, media bias that generated in distinct dissemination mode also obstructs the effectiveness of intercultural communication.

2   Cultural Prejudice and Media Bias Lying in Intercultural Communication

2.1   Cultural Prejudice

As has been discussed frequently, culture has complex definitions and links to numerous disciplinary fields. American anthropologist Geertz [3] defined culture as "an historically transmitted pattern of meanings embodied in symbols, a system of inherited conceptions expressed in symbolic forms by means of which men communicate, perpetuate, and develop their knowledge about and attitudes towards life". Samovar et al. [4] stated culture is the rules for living and functioning in society, in order to function and be effective in a particular culture, people need to know how to "play by the rules", They expounded culture is learned, symbolic, dynamic, transmitted intergenerationally and ethnocentric. If briefly concluded, culture is coined by human but not genuinely existed without human being's reproduction and communication; culture inheritance embodied as symbols is transmitted by media in societal environment; the development and prosperity of culture highly depend on the dissemination of media. Since the complexity of culture, the effectiveness of cultural communication may be affected by numerous factors. While in intercultural communication context, the peculiarity of different cultural identities and cultural beliefs comprise of inevitable factors that lead to cultural prejudice.

Tomlinson [5] claimed that "cultural identity was something people simply had as an undisturbed existential possession, an inheritance, a benefit of traditional long dwelling, of continuity with the past", originating from "culturally sustaining connections between geographical place and cultural experience". Tomlinson [6] further emphasized that although globalization witnessed rapid development of deterritorializing media and communications technologies, "identity is seen as the upsurging power of local culture that offers (albeit multi-form, disorganized and sometimes politically reactionary) resistance to the centrifugal force of capitalist globalization." Once hold rigid cultural belonging sense and firm cultural belief, people are apt to be enmeshed in information cocoon thereby repelling other cultural information, which accordingly hinder the intercultural communication.

2.2 Media Bias

In addition to different cultural identities and cultural beliefs, another factor that affects the communication effect derived from the mode of information dissemination. Since media has been making a constructive contribution by bridging cultural communication, the rapid development of mass media in 20th century is convinced as accelerator of cultural communication. Mass media subverted the rule of audience limited in a certain place or at a certain time in interpersonal communication context, which greatly improved the width and speed of communication. When it comes to an era of globalization, mediated interconnection omits communicative distance, and creates pervasive instant messaging. As is portrayed by Giddens [7], "Globalization can thus be defined as the intensification of worldwide social relations which link distant localities in such a way that local happenings are shaped by events occurring many miles away and vice versa." However, the culture communication is a process emphasizing interaction, thus the single-direction dissemination mode of traditional mass communication which transmits information from one source point to a layer of group audience, inevitably leads to communication failure or cultural discrimination resulting from lack of efficient interaction, especially in a context of intercultural. Even more, in mass media era, media's roles of framing and agenda setting cause media bias. Entman [8] concluded three major meanings of media bias: 1. distortion bias (news that purportedly distorts or falsifies reality); 2. content bias (news that favors one side rather than providing equivalent treatment to both sides in a political conflict); 3. decision-making bias (the motivations and mindsets of journalists who allegedly produce the biased content). Therefore, though mass media shorten the transmission distance and time, the news slant and media bias block the authentic cultural information transmitted to the audience, hence the misunderstanding may obstruct the intercultural communication as a consequence.

Cultural prejudice resulting from different cultural identities and cultural beliefs plays a primary cause of intercultural communication obstacle, whilst media bias caused both by the propaganda mode like agenda setting, as well as ideological concerns from decision-making management of media companies, deepen this obstacle. Since the cultural differences are inevitable, is there any possible to adjust the present communication mechanism, so as to relieve the dilemma of China's culture go-out? This paper tries to tackle this question.

3 Social Media and Advantages of Transmitting Culture

With breakthrough of Web 2.0, the use of social media has become pervasive in people's daily life. The widespread presence of platforms drives people to move many of their social, cultural, and professional activities to these online environments [9]. Social media platforms reconstruct online societal environment without time, space and nationality restrictions, so as to reshape people's cultural communication mode. In that case, will social media break through the dilemma of Chinese culture go-out?

3.1 History of Social Media

Based on Web 2.0 technology, social media offer individuals, communities, and organizations with access to an array of user-centric spaces where they could populate with all kinds of user-generated content [10]. Not only supporting the maintenance of pre-existing social networks, social media also help strangers connect based on shared interests, political views, or activities. In late 1950s, Bruce and Susan Abelson founded a blog equivalence site named "open diary", which enabled people keep online diary in one community, and which originated the initial embryo of social media. Around the year of 2000, a bulk of Social Networking Sites (SNSs) pop out with increasingly growing availability of internet access, allowing users to create profiles, list friends and surf with friends online. Form 2003 onward, newly SNSs launched with social functions such as Myspace, then the term of "social media" began to be prevail by which describing people socially affiliating with others online. In 2004, the most successful SNS Facebook was set up to support niche communities divided by demographics. Facebook pioneered to add customized applications that attract users interacting in tasks, such as comparing movie preferences [11]. Follow the historical roots and technical development of social media, it can be concluded that social media provide open resources sharing, which enable users socially interact through online applications and entail participatory and collaborative commitment. Therefore, Social media are convinced to break through cultural barriers in intercultural contexts since a direct dialogue can be easily established between people with different cultural backgrounds, and escape some possible misleading input conveyed through mass media, no matter by ideological differences or by strong culture invasion.

3.2 Advantages of Social Media When Disseminating Culture

Actually, with regard to the characteristics of openness, interactive and participatory when social media circulating information, traditional media has made great strides in integrating social media in order to make better functions. Researchers from CCTV once revealed the factors that leads to social media transformation: Social media with web2.0 model broke the limitations of old networking communication style and brought convenience to users (technical factor); social media transmit message quicker and more precisely to audience, thereby tremendously arousing citizens' democratic participation (social factor); some soft news that are inappropriate for TV broadcast, yet increasing readership and extra credibility to traditional media when released in social media (political factor); due to increasingly active web users, embracing new media is a good way to retain audience (marketing factor) [12]. Similarly, Social media has following advantages when disseminating culture:

(1) Technological support
     Compared to early static web, Web 2.0 provided a democratized way for people to access and contribute to the Internet [13]. Users were no longer passively reviewing information on the web, but actively generated contents and edited the web page. Thus the direct cultural exchanges may reduce the influence of media bias, and enable culture dissemination impartially in a virtuous circle.
(2) Social platform support
     Different from a single source of information input by mass media (newspapers, books and periodicals) or spectacle of memorial reappearance without audience interactions (TV), social media build immersive platforms that imitate the social

environment, where users continue their social communications in the virtual environment. Ellison et al. conducted a research certified that social media maintain or even solidify users' social connections in offline societal environment [14]. In these social platforms, users maintain the social connections and relationships in real society, and follow the social rules to communicate with others.

(3) De-politicization

It's a basic for providing open service that social media set no strict restriction for users' access. Although some niche communities target users by age or alumni identity with the aim of better promoting community activities among groups with similar community discourses, few social media target users by nationality or ideology. The contents shared by users mostly source from daily life, and focus on personal interest, which deliver a personal oriented cultural information that dispel the binary opposite of strong and weak cultures, oriental and occidental cultures. The issues discussed in social media are usually released by users' emotional expression, or vague personal comments, without any affiliated to political party or cultural orientation. Users spontaneously share, comment or re-post the open resources, but not for political utility.

(4) Marketization

According to CNNIC report, by December 2016, the number of cyber citizens in China has reached up to 731 million, including 666 million IM users, 638 million mobile phone SMS users, 271 million Weibo users and 121 million bloggers [15]. Social media have integrated into people's daily life, and reversed the way of information dissemination. People attain information fast by blog style, and could easily re-post or recreate information in their own blog. Marketing firms call the way of blog-to-blog method of transmitting information viral marketing [16]. With such big group of users and influential power, social media are convinced to have immeasurable potential in transmitting cultural information.

4 The Intercultural Communication Mechanism of Social Media: Form Three Components

Combined with the characteristics of openness, dialogism and participatory, it portrays from three components to discuss the feasibility of social media's new-type communication mechanism, which are Platforms, Users and Contents.

1. Platforms

Social media can be divided into following categories according to different running rules of each platform: collaborative projects (e.g. Wiki), content communities (forums, blogs, video websites as YouTube), social networking sites (e.g. Weibo, Facebook), instant messaging software (e.g. WeChart), and virtual social games (e.g. Second Life). Traditional media make a main function of carrier when transmitting information, whilst social media works not only as carrier, but create immersive environment that provide niche space for communication, constructing unmediated social networks. Additionally, the running rules of specific category of social media greatly affect communication styles of users. Such as SNSs like Friendster and MySpace, which are constructed in a way that requires people to indicate relationships or "friendships" with other participants [16], users consequently obey Friend Mode during communications in these platforms. While in democratic collaborative projects, users follow Collaborative Mode of generating contents. And in the case of instant messaging software, users communicate by Community Discourse Mode [17] in niche communities. The immersive virtual environment of social media platforms aims to promote the perception of users' social presence, which means "degree of salience of the other person in the interaction and the consequent salience of the interpersonal relationships" [18]. Social media establish the unmediated social networks that strengthen the bond of intimacy, and provide synchronous messaging system to support the immediacy of medium. Therefore, the upgrading perception of social presence is convinced to activate users' behavior online, thus users may share information and complete collaborative cooperation more freely and actively.

As being a spatial practice for participatory democracy [19], social media have obvious metrics for intercultural communication, and effects may be doubled if combined with characteristic of each platform. There are numerous examples that can be inspected. When calling for a collaborative memory of unique cultural event or give a definition of cultural terminology, collaborative projects like Wikipedia effectively collect the group of power to locate the answer and keep up a continuous update if required. When it comes to cultural activities that entails both of joint effort and collective binding, SNSs such as Weibo show the charm of strong calling. Agenda-setting topics and activities related to culture if delicately planned, users with the same interests can be appealed to join in and expend the related cultural influence. The instant messaging software like WeChart especially suitable for setting up cultural classes. Actually, live online lecture has become a prevailing function of WeChart. The lecturer upload text, voice and video to control the topic of a discourse group, and the audience may attain answers within permitted time. The newly-born cyber lecture is accessible and efficient to gain cultural knowledge. With regard to knowing about the culture of strange nations faraway in a foreign land, virtual communities provide the best base place. In this respect, online Ethnography research can be carried out without isolation from the distance and time [20].

2. Users

Based on an exploration of online niche communities, Boyd & Ellison claimed that Early public online communities such as Usenet and public discussion forums were structured by topics or according to topical hierarchies, but social network sites are structured as personal (or "egocentric") networks, with the individual at the center of their own community. Thus this kind of SNSs are primarily organized around people, not interests [21]. Due to the egocentric connections, most social media target user by demographics. Therefore, "a large, active, and demographically interesting user base is usually a platform's most precious asset" [22]. The capital asset of social media users consists of several essential: The open demographically information given off online increased social visibility, which made it easier to find each other but also to be scrutinized in public [23]; Social media articulate relationship equivalent to friendship, so if people are Friends online means they're friends on other contexts as well [16]. This friends function are apt to lead to negotiation and cooperation; Social media open the access gate, thereby getting the participation of a large, distributed group with a variety of skills has gotten a lot easier [23]. The old limitations of media have been radically reduced, so the power of discourse return back to the former audience [24]. Users' attention and interests call for prime power of group actions. In sum, social media endow users with flexible tool which match up with social communications, and even make users witness the rise of a brand-new means of collaborative production: (1) The openness of social media allow users reverse the old order of "gather, then share" to "share, then gather". By this spontaneous sharing order, social media are convinced to maintain the sustainable dissemination of cultural information. Users automatically share cultural topics that they're interested, absorbing more peers to follow up and expand the specific topic. The spontaneous gathering of topics avoid the cultural stereotype caused by agenda setting of traditional media, and thus not to quit from intercultural communication activities because of cultural stereotype. (2) The dialogue mode of social media effectively promotes the cooperation among users. Whenever in reality or in virtual world, conversation help people directly know about cultural identity and cultural belief of each other and may revise somewhat negative comments of Chinese culture made by western media. The frequent collaborative production activities in social media reverse the stubborn viewing angle of egocentric cultural principle when dealing with another culture, and adopt conversation and negotiation to improve quality and effect of intercultural communication. (3) The participatory mode of social media vastly increases users' capability of collective action. "Information sharing produces shared awareness among the participants, and collaborative production relies on shared creation, but collective action creates shared responsibility, by tying the user's identity to the identity of the group" [25]. As can be seen, when users with different cultural identities establish a common sense of harmonious cultural world, the shared responsibility is believed as the most powerful social instrument to get rid of the cultural prejudice.

3. Contents

The key characteristic of social media that differentiate form traditional media on content supply is that individuals and communities share, create, discuss, and modify user-generated content in the former, whereas the latter simply expend content: the audience read it, watched it, and they used it as knowledge source bank [26]. Social media employ internet and mobile technologies that enable users generate distinctive contents in the highly interactive context:

(1) Instant sharing

The technological revolution of social tools (mobile phone, SNSs, ect.) reschedule the information release. Anyone who owns a phone or other mobile devices could publish news online that happening around them, and the news almost simultaneously caught by social media users once released. The mass amateurization movement originated from social media shorten or even eliminate the news collection and dissemination time.

(2) Truth squad

The unmediated interconnections of social media escape the ideological interference by media system, users feel more free to carry out discursive protest [27] and reveal the truth of the events.

(3) Comprehensive database

With such large group of users' contributions, social media exhibit a rich variety of information sources. Although the distributions of quality turn out to be highly variant, somewhat even abusive, the rich structure of social media still offer more available data than in other domains [28].

(4) Rich presentation

Integrated expressive form of text, picture, voice and video, the contents of social media take on an unprecedented diversity, in the meanwhile, the fun of generating contents with diversity vastly satisfy user's gratifications, what's more, The richness and interestingness of contents guarantee an optimal reach of online communication. Therefore, cultural information can be circulated in a faster, more authentic, more comprehensive and more rich manner through social media, thereby enabling users uncover the genuine veil of foreign cultures and arouse the interests of the very culture.

On another aspect, the data in social media can be collected and computed, so that users' feedback towards information received can be accessed, which make possible of constructing a feedback mechanism of social media. The characteristic of feedback mode also makes intercultural communication in social media computable. Actually, scholars have already noticed the application of computation both in cultural and social field. In cultural field, cultural computing [29] is taken as a method for cultural translation that uses scientific methods to represent the essential aspects of culture. In social field, social computing describes any type of computing application in which software serves as an intermediary or a focus for a social relation [30]. Thanks to the sweeping progress of big data, the interactions of multiple cultures can be tracking and calculated in data generated in social media when people with different cultures choose to communicate. Consequently, the abstract essentials of cultures could be read and understood through media behavior. Social media open up a new path of intercommunication research, like collecting cultural acceptance data of international audiences, tracking cultural propagation effects and evaluating the effects. Therefore, the establishment of international cultural communication evaluation system is convinced to promote Chinese culture dissemination overseas, thus to raise culture soft power of China.

5   Cross Cultural Hierarchy Phenomenon

According to Van Dijck's survey, the minority core users in social platforms provide a good portion of the content and much of the influence on each platform (10% in Twitter, 4% in YouTube and the similar ratio for Wikipedia). As is disclosed by the theory of Information Cocoons: communications universes in which we hear only what we choose and only what comforts and pleases us [31]. Therefore, after generating by the core users in social media, the diffusion of information relies on the continual passing behavior of the receivers, that is to say, if the receiver actively recreates and shares the information with other users, the information can be still hot in attention, but if the receiver loses the interest of keeping up with the topic, the diffusion of the information stops. It's the simplest rule of information dissemination in social media, while if research focus on how to stimulate the receivers (in this paper described as secondary users and edge users) to keep an active media performance, in other words, to transfer the receivers into core users, the heat of topic may be sustained beyond the expectation.

Based on the theory of acculturation, one direct result of keeping users with different cultural identities and cultural beliefs actively communicate is to break the information cocoon and remove the communication obstacles. In 1936, Redfield et al. [32] pointed out the classic definition of acculturation as: "Those phenomena which result when groups of individuals having different cultures come into continuous first-hand contact, with subsequent changes in the original cultural patterns of either or both groups…under this definition acculturation is to be distinguished from … assimilation, which is at times a phase of acculturation." This explanation tells us that continuous first-hand contact is the precondition of changing the original cultural patterns of communicators. Social media provide virtual societal environment that users form remote transnational places can connect directly, and once users experience cultural changes (the core result of acculturation) like language shifts, religious conversions, and fundamental alterations to value systems [33], the negative effect caused by cultural prejudice and media bias considerably fade away in online intercultural communication. Since social media make good function of making intercultural communication easier and reconciling cultural differences, it is convinced that stimulation measures, such as expanding the group of core users, or activate the secondary users and edge users to sustain the information dissemination, are conducive to intercultural communication.

Actually, a few empirical researches have sketched out the relationships among core users, secondary users and edge users during the information dissemination process in social media [34]. Based on the primary research, this paper coins a layer pattern of intercultural communication. The center circle consists of core users who mainly contribute the contents of one specific culture and actively disseminate information (including share, comment or re-post) related to this culture. The second layer is comprised of those secondary users who might be interest in the target culture but holding other cultural identities and cultural beliefs. They communicate with core users by social media, make changes in cultural habits to acculturate, and help to propagandize the essential of target culture by using the language, sharing the cultural values and to make influence on edge users, who constitute the outer layer of the pattern. Edge users also hold different cultural identities and cultural beliefs with core users, but sometimes come from the same culture with secondary users. They feel unconcerned about target cultural information, and are even not in the least altered by communication related to that culture. The reasons for the edge users' indifference toward target culture may vary, but the fore-mentioned culture prejudice and media bias should be one of those reasons result in their indifference or bias. Therefore, how to swift the stereotype of edge users toward Chinese culture and activate them to join in the group of secondary users to propagandize Chinese culture, which is the key point to break the dilemma of Chinese culture go-out.

Combined with the three components of intercultural communication mechanism of social media, this paper further raises three methods accordingly to activate edge users' media behavior, which is named Cross Cultural Hierarchy Phenomenon (Fig. 1):

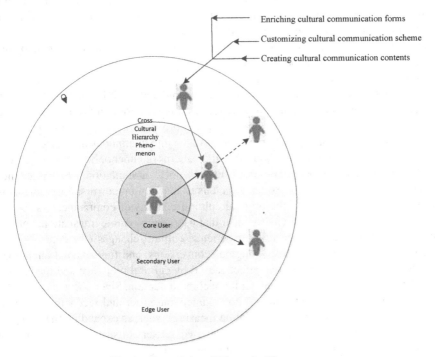

Fig. 1. Cross Cultural Hierarchy Phenomenon

1. Enriching cultural communication forms (utilizing the characteristic of social platform);
2. Customizing cultural communication scheme (by User Typology analysis [35]);
3. Creating cultural communication contents (converging multiple new media technologies).

6 Conclusion

Not only shorten the spatiotemporal distance among users, social media construct online virtual society for users with cultural identities and cultural beliefs to exchange cultural ideas and acculturate in a more flexible environment without political or mediated orientation. The immersive social background, the mode of friendship connections, and spontaneously generated contents endow this type of social tool incomparable advantages to disseminate cultural information and maintain users interest heat of the very culture. As is assumed in this paper, if appropriate measures are adopted to stimulate the users who are indifferent toward Chinese traditional culture, it is convinced that Chinese government locate a new path to walk out of the dilemma of Chinese culture go-out. While this paper only conducts a constructive concept of hypothetical discussion for the Cross Cultural Hierarchy Phenomenon, further research should be carried out to certify this hypothesis with empirical data.

References

1. Xinhuanet: China urges U.S. to abandon prejudice against Confucius Institutes 06 December 2014. http://en.people.cn/n/2014/1206/c90000-8819080.html. Accessed 21 Nov 2019
2. Peiren, S., Jinyun, Y.: Harmony but not uniformity, unity and then success: modern values of excellent traditional Chinese culture. J. Xinjiang Norm. Univ. **36**(6), 52–62 (2015)
3. Geertz, C.: The Interpretation of Cultures: Selected Essays, p. 89. Basic Books Publishers, New York (1973)
4. Samovar, L.A., Porter, R.E., McDaniel, E.R.: Communication Between Cultures, 13th edn, pp. 10–13. Wadsworth Cengage Learning, Boston (2012)
5. Tomlinson, J.: Globalization and cultural identity. In: The Global Transformations Reader: An Introduction to the Globalization Debate, p. 269. Polity Press, Cambridge (2003)
6. Tomlinson, J.: Globalization and cultural identity. In: The Global Transformations Reader: An Introduction to the Globalization Debate, p. 270. Polity Press, Cambridge (2003)
7. Giddens, A.: The Consequences of Modernity, p. 64. Polity Press, Cambridge (1990)
8. Entman, R.M.: Framing bias: media in the distribution of power. J. Commun. **57**(1), 163–173 (2007)
9. Van Dijck, J.: The Culture of Connectivity: A Critical History of Social Media, p. 4. Oxford University Press, New York (2013)
10. Obar, J.A., Wildman, S.: Social media definition and the governance challenge: an introduction to the special issue. Telecommun. Policy **39**(9), 745–750 (2015)
11. Boyd, D.M., Ellison, N.B.: Social network sites: definition, history, and scholarship. J. Comput. Med. Commun. **13**(1), 210–230 (2007)
12. Huang, L., Lu, W.: Functions and roles of social media in media transformation in China: a case study of "@CCTV NEWS". Telematics Inf. **34**, 774–785 (2017)
13. Strickland, J.: How Web 2.0 Works. http://computer.howstuffworks.com/web-202.htm. Accessed 28 Dec 2007
14. Ellison, N., Steinfield, C., Lampe, C.: The benefits of Facebook "friends": exploring the relationship between college students' use of online social networks and social capital. J. Comput. Mediat. Commun. **12**(3), 1143–1168 (2007)
15. CNNIC: The 39th Statistics Report on Internet Network Development of China (2017). http://cnnic.cn/hlwfzyj/hlwxzbg/hlwtjbg/201701/P020170123364672657408.pdf. Accessed 28 Oct 2019
16. Boyd, D.: Friends, friendsters, and top 8: writing community into being on social network sites. First Monday **11**(12), 1–15 (2006)
17. Borg, E.: Discourse community originate from Linguistic theory which means group of people that have goals or purposes, and use communication (a set of discourse, common values and norms) to achieve these goals. Discourse communities. ELT J. **57**(4), 398–400 (2003)
18. Short, J., Williams, E., Christie, B.: The Social Psychology of Telecommunications. Wiley, Hoboken (1976)
19. Shao, P., Wang, Y.: How does social media change Chinese political culture? The formation of fragmentized public sphere. Telematics Inform. **34**(3), 694–704 (2016)

20. Bünyamin, A.: Virtual communities as a social and cultural phenomenon. J. Educ. Learn. **5**(3), 149–158 (2016)
21. Boyd, D., Ellison, N.: Social network sites: definition, history, and scholarship. J. Comput. Mediat. Commun. **13**(1), 210–230 (2010)
22. Van Dijck, J.: The Culture of Connectivity: A Critical History of Social Media, p. 36. Oxford University Press, New York (2013)
23. Shirky, C.: Here Comes Everybody: The Power of Organizing Without Organizations, p. 12. The Penguin Press, New York (2008)
24. Gillmor, D.: We the Media: Grassroots Journalism By the People. For the People. O'Reilly Media Inc, Sebastopol (2006)
25. Shirky, C.: Here Comes Everybody: The Power of Organizing Without Organizations, p. 51. The Penguin Press, New York (2008)
26. Kietzmann, J.H., Hermkens, K., Mccarthy, I.P., Silvestre, B.S.: Social media? Get serious! understanding the functional building blocks of social media. Bus. Horiz. **54**(3), 241–251 (2011)
27. Koopmans, R., Statham, P.: Political claims analysis: integrating protest event and political discourse approaches. Mobil. Int. Q. **4**(2), 203–221 (1999)
28. Agichtein, E., Castillo, C., Donato, D., Gionis, A., Mishne, G.: Finding high-quality content in social media. In: Proceedings of the 2008 International Conference on Web Search and Data Mining, pp. 183–194. ACM (2008)
29. Tosa, N., Matsuoka, S., Ellis, B., Ueda, H., Nakatsu, R.: Cultural Computing with Context-Aware Application: ZENetic Computer. In: Kishino, F., Kitamura, Y., Kato, H., Nagata, N. (eds.) ICEC 2005. LNCS, vol. 3711, pp. 13–23. Springer, Heidelberg (2005). https://doi.org/10.1007/11558651_2
30. Schuler, D.: Social computing. Commun. ACM **37**(1), 28–29 (1994)
31. Sunstein, C.R.: Infotopia: How Many Minds Produce Knowledge, p. 9. Oxford University Press, London (2008)
32. Redfield, R., Linton, R., Herskovits, M.J.: Memorandum for the study of acculturation. Am. Anthropol. **38**, 149–152 (1935)
33. Berry, J.W.: Immigration, acculturation, and adaptation. Appl. Psychol. **46**(1), 5–34 (1997)
34. Haixia, G.: New social networking features and model analysis of information dissemination. J. Mod. Inf. **32**(1), 56–59 (2012)
35. Petter, B.B., Jan, H., Amela, H.: Understanding the new digital divide—a typology of Internet users in Europe. Int. J. Hum. Comput. Stud. **69**(3), 123–138 (2011)

Author Index

Printed in the United States
By Bookmasters